Improving Water Supply Networks

Fit for Purpose Strategies and Technologies

Improving Water Supply Networks

Fit for Purpose Strategies and Technologies

Stuart Hamilton, Bambos Charalambous and Gary Wyeth

IWA PUBLISHING

Published by **IWA Publishing**
Republic – Export Building, 1st Floor
2 Clove Crescent
London E14 2BE, UK
Telephone: +44 (0)20 7654 5500
Fax: +44 (0)20 7654 5555
Email: publications@iwap.co.uk
Web: www.iwapublishing.com

First published 2021

British Library Cataloguing in Publication Data
A CIP catalogue record for this book is available from the British Library

ISBN: 9781780409191 (Paperback)
ISBN: 9781780409207 (eBook)

Contents

Chapter 3

Chapter 4

Chapter 5

Useful Acronyms and Abbreviations

ALC	Active Leakage Control
AMI	Advanced Metering Infrastructure
AMR	Automatic Meter Reading
AOI	Area of Interest
CAD	Computer Aided Design
CAPEX	Capital Expenditures
CAPL	Current Annual volume of Physical Losses
CARL	Current Annual volume of Real Losses
DEM	Digital Elevation Model
DM	District Meter
DMA	District Metered Area
ESPB	Equivalent Service Pipe Burst
FS	Feasibility Study
FTE	Full time equivalent (as in employee)
FY	Fiscal Year
GIS	Geographic Information System
ILI	Infrastructure Leakage Index
IMM	Integrated Meter Management
IRR	Internal Rate of Return
IWA	International Water Association
IWS	Intermittent Water Supply
KPI	Key Performance Indicator
l/conn./d	Litres per service connection per day
LPI	Leak Performance Indicator
m^3	Cubic meter
m^3/a	Cubic meters per year
MAAPL	Minimum Achievable Annual volume of Physical Losses

masl	Metres above sea level
MLD/mld	Million liters per day
MNF	Minimum Night Flow
NPV	Net Present Value
NRW	Non-Revenue Water
OPEX	Operational Expenditures
PI	Performance Indicator
PMP	Pressure Monitoring Point
PRV	Pressure Reducing Valve
UARL	Unavoidable Annual Real Losses
UFM	Ultrasonic Flow Meter
VFD	Variable Frequency Drive
VPN	Virtual Private Network
VSD	Variable Speed Drive
WLSG	Water Loss Specialist Group of the IWA
w.s.p.	When the System is Pressurised (i.e., continuous 24 × 7 supply)
WTP	Water Treatment Plant
ZPT	Zero Pressure Test

Preface

Ageing infrastructure and declining water resources, coupled with a growing global population, are major concerns for water utility operators, and have forced both utilities and users themselves to control water use. The need to manage this valuable resource has pushed utilities to improve their efficiencies, and to adopt various changes in the way they work and operate distribution systems, from bulk water supply to human resource management.

The aim of this book is to provide utility operators with a better understanding of their water supply system, which will form the basis for further progress and system enhancement through better control and management, until eventually the concept of an intelligent system is in place, that is, a self-managed system with minimum human intervention.

Through the Water Loss Specialist Group and its Working Groups, the International Water Association (IWA) has established several relevant guidelines, including the *IWA Standard Water Balance* and the *Basic Management Strategies* for reducing and controlling water losses. The Water System Improvement Matrix proposed in this book was developed based on IWA best practices and methodologies enabling any water utility or water user to understand at what operational level they currently are and what needs to be done to move to the next level.

The Water System Improvement Matrix described here is a practical and simple approach to evaluating the current operational level of a system based on seven Key Areas which are considered crucial in system operations. Each Key Area has three levels of improvement in each of four System Categories, thus building a matrix of twelve improvement steps. A scoring system has been developed for the Improvement Matrix, providing a utility with an overall assessment of their system performance.

Utilities that continuously invest in technology and innovation should see a positive return on investment, in terms of improving daily operations and in the collection and analysis of network data for decision making and forward planning.

Author(s): Stuart Hamilton, Bambos Charalambous and Gary Wyeth
doi: 10.2166/9781780409207_xiii

Methodologies for achieving the best results, to further enhance the capability of a water utility to advance from an intermittent supply system to an intelligent system, are documented in this book and, although the technologies used are continuously evolving, that should not prevent a water utility starting to make advances in their system today.

The continuous development in this area is leading to innovative technologies and new product development to complement current methodologies. The contents of this book reflect the situation at the time of publication.

Part One

Chapters 1–8

Chapter 1

The water system improvement matrix

'There is nothing more difficult to take in hand, more perilous to conduct, or more uncertain in its success, than to take the lead in the introduction of a new order of things.'

Niccolò Machiavelli (The Prince, 1532)

1.1 MATRIX CHARACTERISTICS

For water utilities to move from one operational level to the next may seem an extremely difficult or even an impossible task. Organisational cultures, operational regimes and practices developed over many years in a water utility may act as barriers to moving forward and to achieving higher performance and efficiency. Culture is difficult to observe but over time it becomes organisational tradition which is passed on and taught to new employees as the norm. As an example, intermittency in water supply is considered in many parts of the world as the standard modus operandi in distribution networks.

The use of the Water System Improvement Matrix, described in this chapter, begins at the current level of a water system and then proceeds to the next level through a series of much-needed sustainable changes and system improvements.

It is imperative to fully understand and measure a water utility's current level of operations. Once this level is documented and proven, improvement plans on how to move forward to the next operational level can be developed. It is common for utilities to think that they are indeed better than what they are, thus leading to an underestimation of the effort and work needed to improve system efficiency. It must be understood that to move forward it will take many processes, some simple and some that are not so easy, along with some that will require investment and time.

The Water System Improvement Matrix proposed in this book is a practical and simple approach to evaluating the current operational level of a system, based on seven Key Areas which are considered crucial in system operations, namely:

Author(s): Stuart Hamilton, Bambos Charalambous and Gary Wyeth
doi: 10.2166/9781780409207_0003

(1) Bulk Flow Measurement
(2) Customer Metering
(3) Pressure Management
(4) Leakage Management
(5) Asset Management
(6) Water Balance and Key Performance Indicators (KPIs)
(7) Human Resources

Network operations are broadly divided into the following four system categories:

(A) Basic Network System
(B) Ordinary Network System
(C) Smart Network System
(D) Intelligent Network System

The system categories used in the matrix describe the supply systems that are in existence today. Basic Network Systems are present all over the world with the vast majority located in South Asia, Africa and Latin America. Ordinary Network Systems are the norm in many parts of the world, particularly in the developed countries. In some countries Smart Network Systems have already been employed, having given birth to the concept of 'smart utilities' which employ state-of-the-art technologies and knowhow in their supply system operations.

In addition to the above categories which are known and established in today's water supply systems, we can also set our sights on the future beyond the smart system, and look towards an 'Intelligent' Network System. Today, this may be a dream but perhaps in a few years it could be a reality. Intelligent systems are already revolutionising a variety of industries, including transportation and logistics, security and manufacturing. They help improve the energy efficiency, quality and flexibility of these systems. Intelligent systems are complex and use a wide range of technologies: artificial intelligence, cybersecurity, natural language processing, deep learning, embedded central processing units, distributed storage, wireless networking, etc.

Intelligent systems use artificial intelligence and machine learning, which helps machines to 'learn' in much the same way humans do. This is now possible because of the ubiquitous data in the modern world, and the ability to store it and communicate it at high speeds. We see that in a few years from now there will likely be a breakthrough in utilising such systems in the management of water supply networks. With machine learning, computers take in data and train themselves based on that data. They run tests to ensure they are interpreting the data correctly, and then pass it through classification algorithms to figure out what in the current situation is familiar and what is less well known. When the computers' algorithms have been trained to make the right decisions, they can use this decision-making ability to perform tasks.

Artificial intelligence is definitely the next big step in the operational management of water supply systems, a leap that many people may think is impossible and believe we are dreamers for imagining, but just think how many dreams that were thought to be impossible a few years ago are now possible and pretty much part of our daily lives.

The Water System Improvement Matrix has been developed based on our collective experience in the water utility sector reflecting useful and practical steps for sustainable system improvement. The matrix describes the different levels and steps of each Key Area for four different System Categories. For each Key Area, there are three levels of improvement (with Level 1 being the lowest and Level 3 the highest) in each of the four System Categories, resulting in twelve steps broadly describing the improvement step ladder for that Key Area.

Table 1.1 Scoring system for use in each of the seven key areas of improvement.

Scoring	Basic Network System	Ordinary Network System	Smart Network System	Intelligent Network System
Level 1	1	4	7	10
Level 2	2	5	8	11
Level 3	3	6	9	12

1.2 MATRIX SCORING SYSTEM

The Water System Improvement Matrix can be used to assess a utility's overall system performance based on a scoring system which we have developed. The scoring system shown in Table 1.1 is used to assess the performance in each one of the seven Key Areas.

The score for each Key Area is added to produce a Total Score to assess the Overall System Performance of the utility using the Water System Improvement Matrix we have developed, shown in Table 1.2.

Table 1.2 Water system improvement matrix.

Total Score	Class	Remarks
7–12	E3	Operating procedures failing to make proper use of resources and assets, impacting on the continued deterioration of assets and level of service
13–19	E2	Inferior standard and quality of system management, lacking organised administration and planning of the water supply network, failing to deliver the required level of service and commitment to system improvement
20–26	E1	Critical operational stage, which needs to be bolstered with improved processes, standards and procedures in order to reverse the downward spiral of inefficient operation and take back system control
27–36	D	Satisfactory supply regime but nevertheless wasteful use of resources and inefficient system operation which requires further enhancement and control of system operations leading to increased performance
37–48	C	Efficient use of resources with systems in place that deliver the desired level of service; however, there is room for system efficiencies and optimisation of processes
49–61	B	High degree of efficiency in all system operations, employing advanced technologies and knowhow to sustain the gains of efficiency and to build a platform to enable further advancement
62–73	A	Optimum use of technology, equipment and human resources in operating and maintaining the water supply system in an intelligent, productive, cost efficient and effective manner
74–84	A*	Possibly described now as a 'dream', this would be the supply system of the future operated by machine learning algorithms, self-controlled and managed with minimal supervision by humans

1.3 WATER SYSTEM IMPROVEMENT MATRIX

The matrix provides a classification of water supply systems into eight broad classes (Table 1.2) depicting the different levels of operational efficiencies, with the lowest level describing a wasteful and inefficient system with operating procedures failing to make proper use of resources and assets, impacting on the continued deterioration of assets and level of service. At the top end is a system that could be described, based on today's knowhow and technology, as a 'dream'. This would be the supply system of the future operated by machine learning algorithms, self-controlled and managed with minimal supervision by humans.

Moving forward in system improvement and sustaining the gains of efficiency is hard work, which is even harder with the many problems that water utilities face on a daily basis. Removing some of these problems would make it much easier to move to the next level. Completing a system improvement programme is like climbing a ladder and each step gained means getting closer to the goal.

A successful system improvement programme is not about standing on a moving escalator taking you easily to your ultimate goal but rather about hard work, shedding weight as you go forward enabling you to climb the next rung on the ladder in a steady and secure way.

1.4 KEY AREAS AND IMPROVEMENT STEP LADDERS

A practical and simple approach to evaluating the current operational level of a water system has been devised based on the seven Key Areas which are considered crucial in system operations. These Key Areas are presented in Tables 1.3–1.16 together with their corresponding Improvement Step Ladders and are expanded under each System Category, providing background information on their significance as well as practices, techniques and methodologies to improve them.

Table 1.3 Bulk flow measurement matrix.

Bulk Flow Measurement				
System Level	**Basic Network System**	**Ordinary Network System**	**Smart Network System**	**Intelligent Network System**
Level 1	• No reliable flow measurement • Limited metering of bulk flows in the system	• All bulk flows are metered but not sure about the meter accuracy	• Bulk flows are metered, remotely read, monitored, and analysed with a permanent meter accuracy test programme in place	• Bulk flows metered, remotely read, monitored and analysed with software recognition for meter accuracy drifts, with automatic meter recalibration capability and flow pattern analysis
Level 2	• A continuous programme for the installation of bulk meters	• Meters are manually read and/or remotely monitored with meter accuracy occasionally checked	• Bulk flows are metered, remotely read, monitored and analysed with software to recognise when meter accuracy drifts	• Bulk flows are metered, remotely read, monitored and analysed with software recognition for meter accuracy drifts, with automatic meter recalibration capability and flow forecasting based on historical supply and demand trends
Level 3	• The majority or all of the system bulk flows are metered	• Bulk flows are metered, remotely read, monitored, and analysed with frequent meter accuracy tests	• Bulk flows metered, remotely read, monitored and analysed with software recognition for meter accuracy drifts and flow pattern analysis	• Complete machine learning algorithms that control the system, to manage system input volumes and pressures

Table 1.4 Bulk flow measurement improvement step ladder.

BULK FLOW MEASUREMENT – WATER SYSTEM IMPROVEMENT STEP LADDER											
	BASIC NETWORK				ORDINARY NETWORK			SMART NETWORK			INTELLIGENT NETWORK
Unreliable, limited bulk flow metering	Substantial bulk flow metering	Majority bulk flows metered	All bulk flows metered	Manual reading, occasional meter accuracy checks	Remote reading, frequent meter accuracy tests	Remote reading and analysis, permanent meter accuracy tests	Remote reading and analysis with self-accuracy meter checks	Continuous flow pattern analysis with self-accuracy meter checks	Software controlled with automatic meter recalibration	Software controlled with flow forecasting and automatic meter recalibration	Machine learning algorithms, bulk system control

Table 1.5 Customer metering matrix.

Customer Metering				
System Level	**Basic Network System**	**Ordinary Network System**	**Smart Network System**	**Intelligent Network System**
Level 1	• No or limited customer metering • Unreliable information on the age and type of meters • Customer database has not been updated for a long time	• Substantial or universal customer metering • Customer database is periodically updated • Illegal connections are sporadically detected • Meter readers are occasionally rotated	• Customer database is updated and linked to a GIS • Handheld devices are used for meter reading and bills issued on the spot • AMR meters have been introduced	• All customer meters have sound sensors to detect leak noises, with connected communication systems allowing for automatic correlation and pinpointing of leaks • Fraudulent activities are detected via the AMI and AMR systems
Level 2	• Inadequate meter and customer information • No assessment is made and there is no programme to deal with water theft	• There is a meter replacement programme in place • Customer database is regularly updated • There is a thorough illegal connections detection programme in place	• Demand management programme in place based on consumption patterns identified via a GIS–billing infrastructure • Continuous replacement programme of customer meters to AMR capability	• All customer meters have sound sensors to detect leak noises, with connected communication systems allowing for automatic correlation and pinpointing of leaks • Automatic alerts sent out to customers when excessive customer side usage experienced
Level 3	• Not all customers have meters installed • No regular replacement policy, only when meters stop • No system of controlling meter readers	• All customers are metered with good accuracy meters • Meter readers are rotated, and spot checks are often made • Handheld devices are used for meter reading	• All customers are metered with a high accuracy AMR system • AMI system in place following a strict meter replacement policy	• Complete machine learning algorithms that control the system, so the system is run with internal flow measurement, measuring and adjusting meters for accuracy, with a total self-billing and billed collection procedure without any human interventions

Note: AMI: Advanced Metering Infrastructure; AMR: Automatic Meter Reading; GIS: Geographic Information System.

Table 1.6 Customer metering improvement step ladder.

CUSTOMER METERING – WATER SYSTEM IMPROVEMENT STEP LADDER	
BASIC NETWORK	Limited, poor accuracy meters, unreliable and outdated database
	Inadequate meter and customer database
	Not all customers metered and no meter replacement policy in place
ORDINARY NETWORK	Customers metered, database periodically updated, checks for fraudulent water use
	Meter replacement programme in place, customer database regularly updated, fraud detection programme operational
	All customers metered, good accuracy meters, PDA devices used
SMART NETWORK	Customer database updated and linked to a GIS; AMR and PDA devices in place
	AMR–GIS billing infrastructure and demand management programme
	All customers are metered by a high accuracy AMR system, with AMI system in place
INTELLIGENT NETWORK	Combined customer meters and leak sensors with automatic alerts for fraudulent use
	Combined customer meters and leak sensors with automatic alerts for fraudulent, excessive use and system leaks
	System control via machine learning algorithms with a self-billing and collection procedure

Table 1.7 Pressure management matrix.

Pressure Management				
System Level	**Basic Network System**	**Ordinary Network System**	**Smart Network System**	**Intelligent Network System**
Level 1	• No recording or control of pressure • Perhaps partial network zoning but not used for pressure control	• Permanent pressure measurement at some points with pressure loggers • Some form of further pressure control, e.g. fixed downstream control with a PRV, soft start capability of pumps	• Fully pressure-controlled network • Introduction of pumps installed with VFDs to enable constant pressures with varying demands	• Pressure sensors installed in strategic locations in the network, continuously feeding pressure readings into leak and demand analyses models
Level 2	• Some pressure control, through sizing of zones to maximise system pressures • Pressure monitoring (if any) only at pumping stations and trunk mains	• Pressure zoning in place, e.g. DMAs • Manual analysis of pressure patterns, which are compared to corresponding flow patterns	• Pressure control via PRVs in all areas of the network using advanced types of control, e.g. flow modulation, critical point, etc.	• Multiple pressure sensors installed across the network, analysing system pressures continuously and using data analytics to enable automated control of the pumps and valves to calm the network
Level 3	• Partly pressure controlled through zoning, by elevation • Periodic pressure monitoring using lift and shift pressure loggers	• Pressure control and monitoring at salient points in the network with automatic pressure/flow analysis	• Sensors installed in the network, permanently monitoring pressure variations, which are used to adjust system pressures to develop calm networks	• Complete machine learning algorithms that monitor and optimise pressures in the system, for a completely calm network ensuring maximum asset life

Note: DMA: District metered area; PRV: Pressure reducing valve; VFD: Variable Frequency Drive.

Table 1.8 Pressure management improvement step ladder.

PRESSURE MANAGEMENT – WATER SYSTEM IMPROVEMENT STEP LADDER

BASIC NETWORK | ORDINARY NETWORK | SMART NETWORK | INTELLIGENT NETWORK

1. Rationing zones but no pressure recording and control
2. Network zoning to improve pressure control
3. Network zoning with periodic pressure monitoring
4. Simple pressure control with permanent pressure monitoring at some points
5. Pressure zoning in place and manual analysis of pressure and flow patterns
6. Pressure monitoring and control at salient network points with pressure/flow analysis
7. Pressure controlled network automatically responding to varying demand
8. Fully automated advanced pressure control in all areas of the network
9. Pressure monitoring via network sensors to adjust system pressures to develop calm networks
10. Pressure sensors feeding readings into leak and demand analysis models
11. Multiple pressure sensors in the network with data analytics software enabling automated pressure controlled calm network
12. Complete machine learning algorithms to optimise system pressures ensuring maximum asset life

Table 1.9 Leakage management matrix.

Leakage Management				
System Level	**Basic Network System**	**Ordinary Network System**	**Smart Network System**	**Intelligent Network System**
Level 1	• No leakage control, only repair large mains bursts. No records kept. No ALC programme in place	• Some DMAs are established. Introduction of technologies to assist in ALC activities. Analysis of leak detection and repair records	• Complete sectorisation (DMAs/zoning) in place, and flow and pressure are monitored via permanent installations • Leak detection and repair records are maintained on a GIS platform. • Prioritising leak detection technology based on asset characteristics including ALC on large diameter mains with minimal repair times established	• All data received from permanent monitoring devices and surveys, pulled automatically into a dynamic asset management system, for planning, undertaking and reporting of leakage management and repair activities
Level 2	• Start to undertake a visual ALC programme. Limited leak repair records are kept	• Multiple DMAs established and analyse of leakage and NRW data. Planned regular leak surveys. Short repair times in place. Detailed records of all leaks and repairs maintained and analysed	• Introduction of permanently installed acoustic monitoring systems, with correlating ability. Analysis of system NRW data, undertaken automatically using specialised software. Large diameter surveys undertaken using internal or specialised acoustic technology	• Leak locations automatically detected, pinpointed and linked into a programme to execute the repair process, allowing automated repairs to take place. Repairs undertaken inside the pipe using automated processes, without excavation or water shutdowns, negating the need for customer disruption
Level 3	• Prioritising of ALC activities. Improved leak repair times. Leak detection and repair records kept	• Maximise DMA coverage, with permanent monitoring and communication with a central control room. Leak detection programme prioritised using DMA MNF analysis. • Problematic areas permanently monitored with acoustic devices	• Advanced and innovative technology and equipment used for leakage detection, on all mains diameters and materials, including aerial and satellite applications. Permanent monitoring of system noises and pressures used to analyse changes in historical data and predict new leaks/bursts	• Complete machine learning algorithms that control the system and undertake analysis of the network to identify asset weaknesses and potential leak locations, analyses causes of failure and where possible rectifies the cause, whilst linked into the asset management system to prioritise systematic pipe replacement, including the optimum choice of pipe material

Note: ALC: Active Leakage Control; DMA: District Metered Area; GIS: Geographic Information System; MNF: Minimum Night Flow; NRW: Non-Revenue Water.

Table 1.10 Leakage management improvement step ladder.

LEAKAGE MANAGEMENT – WATER SYSTEM IMPROVEMENT STEP LADDER
BASIC NETWORK
No ALC in place, repair visible leaks on mains and connections only
Visual ALC programme, limited leak repair records kept
ALC activities prioritisation, improved leak repair times, leak records kept
ORDINARY NETWORK
Some DMAs established, ALC technologies introduced, leak records analysed
Multiple DMAs established, regular leak surveys, short repair times in place
DMAs maximised, leak detection prioritised using DMA MNF, permanent acoustic devices in problematic areas
SMART NETWORK
Complete sectorisation, leak records and analyses GIS platform, large diameter mains ALC, minimal leak repair times
Permanently installed self-correlating acoustic systems, large diameter surveys using internal or specialised acoustic technology
Advanced and innovative aerial and satellite technology used for leakage detection on all mains diameters and materials
INTELLIGENT NETWORK
Data pulled automatically into a dynamic asset management system allowing proactive operational control
Leak locations automatically detected, automatic pipe repairs undertaken using in-pipe processes, negating need for supply disruption
Complete machine learning algorithms controlling system, network analysis to identify potential leak locations, asset failure analyses, optimum choice of pipe replacement material

Note: ALC: Active Leakage Control; DMA: District Metered Area; GIS: Geographic Information System; MNF: Minimum Night Flow

Table 1.11 Asset management matrix.

Asset Management				
System Level	**Basic Network System**	**Ordinary Network System**	**Smart Network System**	**Intelligent Network System**
Level 1	• Extremely poor asset condition • No management of or investment in asset infrastructure	• Reasonable estimates of asset renewal requirements • Planned asset management programme developed and operational	• Excellent skills in asset repair – repairs undertaken quickly to minimise water loss • Detailed records of asset maintenance are kept that indicate location, type, date and duration of repair, and are linked to a digital asset management system	• Target replacement and rehabilitation of assets based on actual network performance parameters which will be permanently and continuously monitored through appropriate sensors, such as failure frequency, loss of pressure, reduction in flows, etc.
Level 2	• Limited abilities and capacities to repair critical assets • Basic skills for network maintenance – often long delays for repair and quality of repairs is a problem • Limited planning and maintenance of critical assets	• Very good skills and commitment to asset repair, only occasionally have ongoing problems • Long term asset management plan developed and approved, with funding available to deliver • Detailed records of asset maintenance are kept that indicate location, type date and duration of repair	• Based on good pipe performance and maintenance history, combined with appropriate forecasting techniques, with a well-defined asset renewal strategy including timing, costing, operations and impact on service delivery • Developed repair or replace asset prioritisation system	• New pipe installations of new material, capable of detecting leaks and self-repairing • Introduction of permanent sensors on pipelines for monitoring asset condition, including life expectancy
Level 3	• Reasonable abilities and capacities to repair and replace assets. Minimal capital expenditure budget available • System in very poor condition, therefore asset renewal or improvements reliant on significant funding	• Development of a digital asset management system, with mapping and database capabilities • Risk analysis and management of assets to minimise critical failures • Details of failure analysis documented and characterised	• Analyses asset condition and remaining asset life, by use of advanced technologies • Utilises pipe rehabilitation methods, rather than full pipe replacement to minimise disruption and reduce costs	• Complete machine learning algorithms that control the system and undertake analysis of the network to identify asset weaknesses and potential leak locations, analyse causes of failure and where possible rectifies the cause, whilst linked into the asset management system to prioritise systematic pipe replacement, including the optimum choice of pipe material

Table 1.12 Asset management improvement step ladder.

ASSET MANAGEMENT – WATER SYSTEM IMPROVEMENT STEP LADDER
BASIC NETWORK
No asset management, limited repair of critical assets, no investment in asset infrastructure
Basic skills for network maintenance, limited critical assets planning
Reasonable asset repair and replacement, asset improvement reliant on significant funding
ORDINARY NETWORK
Planned asset management programme developed and operational
Long term asset management plan, funding available to deliver, maintenance records kept
Development of a digital asset management system with mapping and database capabilities, asset risk analysis and management
SMART NETWORK
Detailed records of asset maintenance kept and linked to a digital asset management system
Appropriate forecasting techniques, well-defined asset prioritisation renewal strategy
Advanced asset condition assessment, innovative rehabilitation techniques and materials
INTELLIGENT NETWORK
Appropriate sensors for permanent asset performance monitoring detecting assets failure
Self-leak detection and repair pipes, permanent asset monitoring sensors on pipelines
Complete machine learning algorithms controlling the system, asset failure analysis and rectification, optimisation of asset replacement

Table 1.13 Water balance and KPIs matrix.

Water Balance and KPIs				
System Level	**Basic Network System**	**Ordinary Network System**	**Smart Network System**	**Intelligent Network System**
Level 1	• No water balance established	• Annual water balance in accordance with the international (IWA) format • Regularly calculates physical and commercial loss performance indicators • Occasionally calculate KPIs such as ILI	• Water balance updated regularly, with latest billing and flow data and used to prioritised NRW activities	• Water balance and Minimum Night Flow analyses performed automatically per DMA/zone relevant software and intervention activities prioritised on a daily basis
Level 2	• Attempts to establish water balance using water utility's own water accounting methodology • Only KPI used is %NRW based on water utility's own water accounting methodology	• Establish an annual water balance in accordance with the international (IWA) format and use of 95% confidence limits to indicate accuracy bands	• DMA water balance automatically updated daily, utilising AMI/AMR data and DMA flow data	• All data and information to develop a water balance and KPIs are pulled in automatically from respective databases, to be analysed automatically and improvement actions prioritised
Level 3	• Attempts to calculate NRW performance indicators other than percentage	• Regularly calculate physical and commercial loss performance indicators and publish them in our annual report. • Use KPIs such as ILI for benchmarking	• Water balance and relevant KPIs calculated automatically using relevant software linked to bulk flow measurements, billing and asset management databases	• Water balance calculated daily complete with accurate KPIs with full financial costings, calculated with daily costs for chemical and direct costs calculated from access to the financial systems and online chemical costs. Errors calculated and adjusted based on machine learning algorithms

Note: AMI: Advanced Metering Infrastructure; AMR: Automatic Meter Reading; ILI: Infrastructure Leakage Index; IWA: International Water Association; KPI: Key Performance Indicator; NRW: Non-Revenue Water.

Table 1.14 Water balance and KPIs improvement step ladder.

WATER BALANCE AND KPIs – WATER SYSTEM IMPROVEMENT STEP LADDER											
	BASIC NETWORK			ORDINARY NETWORK			SMART NETWORK			INTELLIGENT NETWORK	
No water balance established	Attempts to establish water balance, % NRW is the only PI used	Attempts to calculate NRW PIs other than %	IWA annual water balance, limited KPIs calculated	IWA annual water balance with 95% confidence limit, some KPIs calculated	Real and apparent loss indicators regularly calculated, KPIs used for benchmarking	Water balance updated regularly, used to prioritised NRW activities	DMA water balance automatically updated daily, utilising AMI/AMR system	Water balance and relevant KPIs calculated automatically using relevant software	Water balance and Minimum Night Flow analyses performed automatically prioritising intervention activities	All relevant data and information pulled in automatically to be analysed and improvement actions prioritised	Water balance calculated daily complete with accurate KPIs with full financial costings based on machine learning algorithms

Table 1.15 Human resources matrix.

Human Resources				
System Level	**Basic Network System**	**Ordinary Network System**	**Smart Network System**	**Intelligent Network System**
Level 1	• No staff training or education and no related budget • No measurable efforts in NRW management	• Staff training and capacity building, availability for an education plan in operational and maintenance activities	• Staff training programmes for all new technologies and systems	• Continued professional development of staff to improve and build knowledge and capacity in intelligent systems
Level 2	• Basic training for some operational field activities provided, mostly on-the-job training • Training efforts in NRW management provided, but mostly opportunistic	• A cross-departmental NRW unit in place dealing efficiently with both real and apparent losses, with adequate staff to undertake the activities	• Improved staff development programme in place addressing cutting edge technologies and systems	• Coordination with research institutions in areas of system improvement using artificial intelligence and systems
Level 3	• Coordination between technical and commercial departments is being introduced as part of efforts to reduce NRW • Sustained and adequate staffing levels to deliver planned programmes	• Actively managed staff training and capacity building, based on a comprehensive and budgeted plan	• Appropriate staffing levels and capabilities, to complement the advanced and innovative technologies in place	• Match staff levels and capabilities to complement intelligent systems adopted

Note: NRW: Non-Revenue Water.

Table 1.16 Human resources improvement step ladder

HUMAN RESOURCES – WATER SYSTEM IMPROVEMENT STEP LADDER	
BASIC NETWORK	No staff training or education and no related budget
	Basic training for some activities, mostly on-the-job training
	Utility departments introduced in NRW reduction efforts
ORDINARY NETWORK	Educational plan in O&M activities in place for NRW reduction
	An NRW unit in place with adequate staff to undertake relevant activities
	Actively managed staff training and capacity building based on a comprehensive and budgeted plan
SMART NETWORK	Staff training programmes for all advanced technologies and systems in place
	Improved staff development schemes in place on cutting edge technologies and systems
	Appropriate staffing levels and capabilities to complement the advanced and innovative systems in place
INTELLIGENT NETWORK	Continued professional development of staff to improve and build knowledge and capacity in intelligent systems
	Work with research institutions in using artificial intelligence for system improvement
	Match staff levels and capabilities to complement intelligent systems adopted

Chapter 2

Water distribution systems

A distribution system is used to describe collectively the facilities that are utilised to supply water from its source to the point of usage.

The primary purpose of a distribution system is to deliver wholesome and safe drinking water to all properties that are connected to the system at an adequate pressure and flow to allow the household or commercial premises to function.

2.1 REQUIREMENTS OF A FUNCTIONAL DISTRIBUTION SYSTEM

A water distribution network is a system of pipes providing appropriate quality and quantity of water to users. The construction and layout of networks have to be carefully prepared in order to guarantee sufficient pressure and ensure hygienically safe water. Good operation and maintenance (O&M) – including repair, leakage control, preventing recontamination, etc. – have to be ensured. Water quantity and quality should not be compromised by the infrastructure system.

There are several types of distribution system; four are described below:

- dead end or tree systems;
- grid system
- circular or ring systems;
- radial systems.

The key requirements of an adequate distribution system are summarised as follows:

- it should be capable of supplying water at a sufficient pressure;
- it should be capable of delivering water for firefighting purpose;
- the design of the system should be such that maintenance can be completed without causing the system to be turned off as a whole and so restricting those who may not need to be affected by such work;

Author(s): Stuart Hamilton, Bambos Charalambous and Gary Wyeth
doi: 10.2166/9781780409207_0021

- installation should be to local byelaws and to a depth where frost or heat cannot affect the pipeline;
- it should be watertight so as not to leak on installation and be capable of lasting about 50 years;
- water quality should not deteriorate while in the distribution pipes.

Operation and maintenance of a water distribution system describes all the activities needed to run the system continuously to provide the necessary service. The objective of operation and maintenance is to ensure an efficient, effective and sustainable system that delivers the required level of service to all users at all times.

Non-Revenue Water (NRW) management of a distribution system is of the utmost importance and is used as a proxy indicator for efficient network operation. There are several basic elements that need to be reviewed and addressed in a NRW management strategy which are in line with the IWA NRW Water Balance:

- bulk flow management;
- customer metering;
- pressure management;
- leakage management;
- asset management;
- water balance and KPIs;
- human resources.

All of the above elements need to be reviewed by a water utility as it moves up the Water System Improvement Step Ladder, since improvements to all or some of them may need to be made to take the next step up.

As with any company, a water utility has income from customer revenues, and outgoings as operational and capital expenditure. Most companies need to work at a profit to ensure continuation of business, but unfortunately that is not the case with many water utilities.

The supply of water has become a very politically motivated business, with increases in tariffs often resulting in failure due to political interventions. To this end many utilities have to work with very low tariffs, which mean that they make a commercial loss every year. This in turn means that they have to survive on subsidies from local or federal governments. Many improvement projects, such as asset replacement and meter replacement, often need to go through a government procurement process, where approval is never guaranteed.

In the long term it is essential that all water companies aim to become financially independent and responsible for their own operational and capital budgets. This can be done through a number of means, such as tariff restructuring, NRW reduction and customer meter replacement. This will often mean that an injection of cash is needed to start the process, either from the government or through some form of performance based or private financing scheme.

For a water utility to become self-sufficient, it does not necessarily mean that it has to be privatised; it can also be achieved whilst remaining a publicly owned utility functioning in a technically and financially efficient and effective manner.

2.2 BULK FLOW MANAGEMENT

The main function of a bulk flow meter is to measure the water flow and record the total volumes that pass through the meter. Bulk flow meters are strategically located along the water network and allow measurement and monitoring of the volume of water produced from sources, conveyed to water treatment plants, and from there to service storage reservoirs from which water is distributed to the various districts of the network before reaching the customers.

Bulk flow measurement should be differentiated from System Input Volume (SIV) since bulk flow refers to all volumes of water that are measured before the water enters the distribution network in order to reach consumers. In this context the measurement of flows at the source, treatment plant, reservoir and at the head of the distribution network, e.g. at a District Metered Area (DMA) inlet, are all characterised as bulk flow measurements. SIV, on the other hand, is the volume of treated water input to that part of the water supply network to which the water balance calculation relates.

NRW analysis begins with calculation of the IWA Water Balance, a key input of which is SIV. This is the volume of water that enters a system, generally from a water treatment plant (WTP) but sometimes as a transfer from another supply system.

Often the number of inputs to a system is relatively low, depending on how many WTPs and transfer connections there are. It is therefore essential that the measurement of these inputs is continuous and done as accurately as possible. Any errors at this point in the NRW calculation will be compounded throughout the IWA Water Balance giving inaccurate estimates of NRW at the end.

As there are a low number of inputs, only a few meters are required, although sometimes these may be of large size. But the low numbers means that higher quality and more accurate meters can be sourced and installed.

2.3 CUSTOMER METERING

One of the other key inputs into the IWA Water Balance is customer meter data, either as Billed Metered or Unbilled Metered. In this instance, a utility is generally dealing with a large number of meters, depending on the number of customers in the system. The cost of these meters is substantial and normally water utilities tend to use cheaper meters for this purpose, and often replacement of these meters is delayed beyond their accurate life span or sometimes not until they fail. This has a double impact on the water utility, as it affects their overall NRW levels but also has a direct impact on their revenues.

These apparent losses (sometimes known as commercial losses), are nearly always less in volume than real or physical losses; however, this does not mean that they are less important. In fact, despite being less in volumetric terms, in monetary terms they are usually higher in value. In addition, reducing apparent losses is often easier and quicker to achieve than reducing real losses, and will have an immediate impact on the revenue received by the water company. Apparent/commercial losses mainly consist of the following:

- customer meter under-registration;
- illegal connections and all other forms of water theft;
- problems and errors in metering, data handling and billing.

Minimising customer meter under-registration was once considered a very difficult task to deliver. However, significant research has been carried out in recent years, the results of which have made the identification of the process and delivery of metering improvement programmes easier to achieve. It is imperative that utilities select appropriate meters for their specific types of installation design and layout, as this is a major cause of meter under registration along with choosing poor design and low-cost meters. Water meters can be considered as a cash generating tool and any reduction of flows measured, due to meter malfunctioning or under-registration, means that revenue is being lost. It is for this reason that the cost of purchasing customer meters should not be the primary criterion for choosing them.

Durability and performance of meters is important. Although all meters are verified to meet specifications by their manufacturers, once installed in the field under working situations, they may deteriorate very quickly. In order to avoid loss of revenue, it is important to establish an optimum period for meter replacement through a structured accuracy testing programme of existing customer meters. Such a programme should not only be limited to small domestic meters but also include larger commercial meters, as these meters measure larger volumes of water and any under-registration would mean higher loss of revenue.

For this reason, it should be considered worthwhile investing in high quality and accuracy meters for large customers. These meters normally yield substantial income for a water utility so an investment in modern and more accurate meters can be justified. Recent advances in technology make it possible, at an affordable cost, for such large meters to be monitored by smart loggers, where stored data is transferred to the utility by means of a cellular network, thus being able to quickly identify any abnormalities, such as fraud action, low/high flows, etc.

2.4 PRESSURE MANAGEMENT

In a water distribution network, pressure is the energy that pushes water through the pipes. Also, this water pressure determines how much water may come out from household taps. The amount of pressure at a household tap is also determined by where the property is located compared to the source of water, which could be supplied to the network, for example, from a ground storage reservoir, elevated water tower or pumping station.

Low pressure can reduce water flow to a slow trickle or even to no water at times, resulting in customers filling sinks or pots to store water for later use. People may also start to install large storage tanks to hold a day's supply or more, or even worse install a suction pump which draws directly from the supply pipe. Where customers utilise water directly from the mains, low pressures can affect customer demand and, in some cases, household devices such as electric showers can also be affected. There are many reasons that can cause low pressure and the majority of these are not the fault of the homeowner but a problem in the distribution system, such as:

- inadequate pumping facilities;
- water mains that are too small or blocked by encrustation; or
- high leakage levels.

On the other hand, pressure that is too high can cause issues such as damage to fittings, pipes and household items. In addition, higher pressures have the effect of increasing leakage volumes, by forcing more water out through the existing leaks. This is why many water utilities try to reduce their water pressures as much as possible to limit the leakage volumes, although reducing pressure actually makes active leak detection more difficult, as will be discussed later. The main causes of high-water pressure are:

- property at a low elevation, e.g. at the bottom of a hill when the source of water is on a hilltop;
- high pumping head;
- transient pressure from a poorly managed system.

In reality the management of system pressures is a balancing act, in which the aim of a utility is to reduce high pressures (in order to control leakage volumes and reduce burst frequency) whilst trying to ensure an adequate level of service, i.e. pressure and flow, to customers at maximum demand.

An approach that is widely used throughout the world is to pressure manage the inflow to DMAs. A pressure reducing valves can be installed at the inlet point so that pressure can be controlled into the area which in turn reduces the number of bursts that occur and, if a burst does occur, the volume of losses is greatly reduced. There is no ideal size for a DMA. Whether it is ultimately 500 or 5,000 service connections is decided on a case-by-case basis and depends on a number of factors (e.g., hydraulics, topographical features, number of boundary valves, practicality of design and installation, along with economic set up costs and benefits).

The size of a DMA has an impact on the cost of creating it. The smaller the DMA, the higher the cost. This is because more valves and flow meters are required, and maintenance is costlier. Costs and issues with boundary valves are removed in the case of a DMA that has natural boundaries.

2.5 LEAKAGE MANAGEMENT

Many water utilities only conduct visual leakage repair. In reality, this is not sufficient to maintain a good working system since as many as 90% of leaks can be below the surface. Adopting operational systems such as those which monitor the minimum night flow in a DMA or by having a smart water network will assist leakage teams to find leaks soon after they occur. A traditional approach for leakage teams to adopt is to listen and check all available fittings on a pipeline by various means, such as simple mechanical listening sticks, electronic ground microphones or even leak noise correlators.

A leakage programme surveying a network once a year would result in an average runtime of a leak of 6 months. If this intervention period is shortened, then the leak run time would be reduced. However, this blanket approach is only satisfactory provided that the network is in reasonably good condition and the natural rate of rise (NRR) of leaks is reasonably low. It is evident that this approach is not targeted at minimising leak run time resulting in high leakage, particularly in networks in poor condition. An improved approach is to split the network into hydraulically discrete zones – DMAs – which are hydraulically discrete and ideally have only a single inflow point. The inflow and corresponding pressure are measured and monitored on a continuous basis. Ideally, when the entire distribution network is split into DMAs, the utility has several advantages:

- the volume of real losses (leaks) can be analysed on a daily, weekly or monthly basis;
- leak detection works can be prioritised;
- new bursts can be identified immediately by monitoring the minimum night flows;
- awareness and location time will be reduced.

2.6 ASSET MANAGEMENT

Providing sufficient water of appropriate quality and quantity is the most important requirement of a network today and one that every water company strives to achieve, as has been the case in history going back thousands of years. In the days of the Roman empire, water was distributed around houses and other properties by a form of piped system, and this could be said to be the birth of the modern water distribution system.

All water systems consist of the same basic characteristics: a source where water is collected from and a storage facility where it is held until required. The water is then distributed through an infrastructure system to where it is used by customers, whether domestic, commercial or industrial. It is the aim of all water suppliers to maintain a 24 × 7 water supply; however, this may not be the case in many instances and the reasons for failure to do so are many.

The infrastructure system of a water distribution system includes pipes, valves, pumps, reservoirs, etc., all of which have a definitive life span. As each asset progresses through its life, it ages and may start to fail, which ultimately could have an impact on the level of water losses. Water utilities should embark on a comprehensive maintenance plan to ensure that the maximum lifespan for each asset is achieved and that water losses are minimised. However, no matter how good the maintenance plan is, assets will eventually reach the end of their effective life and will need to be replaced. This asset replacement also needs to be planned carefully, especially with those assets of huge volume or high cost, to ensure the budget is spread over many years and not left until it is too late and impossible to finance.

2.7 WATER BALANCE AND KPIs

The main purpose of a water balance is to establish a baseline reference for the NRW situation in a network. It is important, however, to complement this with an assessment of performance and identification of the areas of improvement.

A significant contribution to reaching the point of water accountability was the establishment of the IWA Water Balance, which is a useful tool in analysing the various components of water production, storage and distribution. Through this analysis, a water utility can gain an understanding of the magnitude of their water loss problem and set priorities for rectifying the situation, based on component analysis of the identified revenue and non-revenue water elements.

After performing an initial 'top down' water balance, it may become evident that some of the numbers gathered are approximate estimates and inspire little confidence in their accuracy. The next action in the water audit process is to refine the quantities that may have been initially estimated. A 'bottom up' approach is often implemented after the 'top down' audit has been completed, which can help in identifying the real losses component more accurately, thus adjusting the initial apparent losses estimate. A 'bottom up' approach begins by looking at components or discrete areas (DMAs) in the distribution system. It also assesses and verifies the accuracy of the water loss data associated with individual components of the distribution system.

The water balance is an effective tool available to quantify water consumption and losses that occur in the distribution system and in the management processes of the utility. The auditing process provides great insight into the nature and magnitude of water losses occurring in the utility and also helps the utility staff to focus on the practices needed to control water losses.

Furthermore, a benchmarking process involving the evaluation of internationally accepted performance indicators (PIs) enables system performance to be monitored and assessed. The adoption of a sound performance indicator system is essential for improving performance through a sustainable increase in efficiency and effectiveness of operations, as well as an improved quality of service.

A system performance assessment using appropriate indicators is a useful tool in establishing and monitoring progress and benchmarking. The purpose of benchmarking is to search for and identify best practice with the objective of implementing appropriate best practice and improving performance. Data collection is an integral step in the benchmarking path to improved performance.

2.8 HUMAN RESOURCES

Human resources are one of the most important assets of any water utility, and are responsible for assuring systems management and improvement. Strategic management should include the long-term planning of human resources alongside urban water infrastructure assets, to ensure service sustainability. A stable human resources framework and its capacity to efficiently and effectively manage system operations should be assured to maintain the management of an extensive, diverse infrastructure with long life cycles, providing an adequate level of customer service and acceptable risk levels, capacity building and knowledge transfer. Strategic management should include the long-term planning of human resources alongside urban water infrastructure assets, to ensure service sustainability.

Specialised and dedicated ongoing training of the different levels of staff is an integral part of the whole staffing process. Concepts such as accountability of staff through proper structuring and target setting are worthwhile introducing. Capacity building should include an analysis of human resources:

- systematic institutional analysis should be undertaken before a decision is made on the areas and direction of change;
- a short- and medium-term strategy for capacity building should be developed with interim milestones;
- on the basis of the above strategy, training programmes and workshops should be held for all staff, from senior staff right down to field staff, which of course should not exclude anyone;
- through the above, all staff will be made aware of the vision, goals and strategy of the NRW division, and will be trained in order to improve their skills and reinforce their commitments to achieve these.

A capacity development strategy should be based on a systematic institutional diagnosis, including:

- external parameters (e.g., water resources and their management, stakeholders and political pressures, and legal context);
- internal management structures and practices (strategic goals, commercial aims, financial management, structure, reporting channels, and human resources including policies, issues and incentives);
- operations (client focus, investment choices, service delivery, operation and maintenance programmes, leak detection and other factors).

Chapter 3

Basic network systems

In many Basic Network Systems the infrastructure is under-developed to such an extent that it has to be run as an intermittent system. There are also cases where a Basic Network System is run under a continuous supply regime but still has many of the operational issues that an intermittent supply system does. This chapter therefore deals with supply systems that are currently operating on an intermittent basis.

With an intermittent water supply (IWS) system, the water utility's main aim is to ensure that all customers receive water for a variable amount of time. This can be a few hours a day or, in some cases, a few days per week. The water companies are not necessarily concerned about water losses and, consequently, the basic measures enabling the assessment, reduction and control of water losses are not put in place.

The operation of networks under IWS leads to various operational problems that reduce the efficiency of water distribution and use for both water utilities and consumers. Contrary to the intention of extending water resources that drives utilities to initially attempt IWS, it has now been recognised that IWS leads to greater quantities of water being lost. On the consumer side, and in comparison to continuous supply, water supply under IWS conditions leads to excessive consumption as stored water tends to be discarded by consumers when the new supply comes in. Additionally, private storage tanks often overflow and taps that are left open lead to uncontrolled water loss.

On the utility side, the operation of IWS leads to ineffective supply and demand management, with inaccurate customer meters damaged by the frequent emptying and filling of the network and the subsequent vacuum and excessive air conditions in pipes. Moreover, IWS results in inefficient operation with direct financial burdens on the water utility. This includes a decrease in revenue due to decrease in water sales and willingness to pay, as well as the additional expense for staff needed for frequently opening and closing valves, and for conducting repairs to the increased number of bursts.

Author(s): Stuart Hamilton, Bambos Charalambous and Gary Wyeth
doi: 10.2166/9781780409207_0029

3.1 BULK FLOW MANAGEMENT

All water meters require the pipe to be full in order to accurately measure flow rate. Unfortunately, under an intermittent supply system the pipework is often empty or operating at partially full, which tempts many utilities not to install flow meters on main supply pipes, which in turn means that NRW calculations cannot be accurately performed.

However, generally with IWS systems, the transmission or larger diameter network is normally continuously fully pressurised. Flow to the distribution system, by a connection, from the transmission mains (or from a reservoir outlet in the transmission system) is controlled by the opening and closing of isolation valves to intermittently feed distribution areas. In such cases, it is possible to install a flow meter in the transmission system before the isolation valve. This ensures the flow meter is always full of water and measuring full flow when the isolation valve is open.

It is important that any flow meter is installed to the manufacturer's specifications, specifically ensuring the necessary straight lengths of pipe before and after the meter. In addition, opening the isolation valve must be done slowly to enable air to escape sufficiently, to reduce the possibility of transient damage to the pipes and to ensure there is enough back pressure behind the isolation valve to ensure the flow meter remains full.

If the intermittent supply is managed consistently, with the same opening and closing times used daily, then an automatic valve system could be implemented. This would involve some form of actuator with a time control system to open and close the valve slowly ensuring less disruption to the system.

3.2 CUSTOMER METERING

Intermittent water flow in supply and distribution networks may result in a deterioration of water meter accuracy, ultimately leading to malfunction due to high water velocities developed during the filling of the pipelines, and due to the transport of suspended solids in the system and the expulsion of air through the meters, thus seriously damaging the registration mechanism. This situation affects most water meters installed in distribution networks, whether they are used to measure bulk water flows or customer consumption. These conditions cause metering malfunction errors due to the following reasons:

- the alternation of dry and wet conditions weakens the metering devices;
- when air is driven out of the pipes by incoming water, the air movement accelerates the deterioration of the meters' recording mechanisms;
- when water is cut off, negative pressures create vacuum conditions in the pipe system, reversing the meter reading counter.

These factors can also contribute to make consumers suspicious of the accuracy of their bills and less willing to pay, which can sometimes lead to criminal behaviour – however, it should be noted that this is more or less pronounced, depending on the cultural sphere concerned and does not necessarily depend on intermittency, even if it may be aggravated or justified by the feelings of dissatisfaction and injustice that IWS generates.

All of these problems lead water utilities to question the relevance of installing meters for this type of supply, but their absence generates other complications such as the obligation to introduce flat-rate or roughly evaluated billing, and the difficulty of estimating network performance.

3.2.1 Water wastage

Consumers in intermittent conditions may waste more water than those who receive a permanent supply. In fact, they tend to keep their taps open to store as much water as possible, from fear of shortages, which incidentally causes occasional overflows in tanks. In addition, most consumers do not use all the water stored, and this water is thrown away and replaced by fresh water from the next supply window, especially when water tariffs are flat rate or state-subsidised.

This situation is all the more paradoxical as intermittency is often initially caused by a lack of resources. This phenomenon is a salient example of the interweaving between the causes and consequences of intermittency, and the difficulty of improving supply conditions in the next supply window, especially when water tariffs are flat rate or state-subsidised.

These problems lead water utilities to question the relevance of installing and/or replacing stopped or damaged meters for this type of supply, but their absence generates other complications such as the obligation to introduce flat-rate or roughly evaluated billing.

3.3 PRESSURE MANAGEMENT

In an intermittent water supply system, the main aim is to supply water to people as quickly as possible. Generally, this means that customers have taps open continuously, as they want to fill up storage tanks and buckets as fully and as quickly as possible, before the supply goes off. Therefore, pressure management in an intermittent water supply system generally involves turning the water supply on and off at timed intervals. This turning on and off is often achieved by opening and closing a gate/isolation valve, or in some cases by turning a booster pump on and off.

In a managed intermittent supply system, the water supply company generally has enough water supply for customers' needs but does not have the operational capability to turn the supply system into a continuous one. However, when the water supply is first turned on, the flow rate and consumption will be high, with relatively low pressure. But over time, customer tanks fill up and consumption drops, and thus pressure in the pipelines starts to rise. In these instances, it is worth having some form of pressure control to ensure that the system does not get over pressurised at any point in time. If the supply timing is controlled by either gate valves or a booster pump, then the installation of a Pressure Reducing Valve (PRV) could help with limiting any over pressurisation. PRVs can actually close water tight, effectively stopping the water supply, so could be used to turn on and off the supply through either manual control or through a simple battery powered pre-set timer.

Some intermittent water supply systems are quite advanced in the way that water is fed to the distribution system. For example, one water authority in Maharashtra, India, has a major trunk main with some 60 m pressure, that has a number of branches off it feeding individual villages. These villages receive water for a fixed number of hours each day, except for one day a week when the system is totally shut off for maintenance purposes. The individual branches to each village are controlled using an automated PRV. A battery powered timer opens and closes each PRV for a fixed number of hours per day. In addition, the PRVs have a mechanical device installed that limits the flow rate through the valve. This means the PRV system supplies water over a fixed period of time, at a fixed pressure and at a fixed flow rate, therefore fixing the volume of water supplied to each village.

If an intermittent system is supplied by a booster pump, then advanced pumping systems including variable speed drives can be utilised to control flow and pressure in a similar way.

3.4 LEAKAGE MANAGEMENT

3.4.1 Leakage in an Intermittent Water System

There are many difficulties when conducting Active Leakage Control (ALC) in areas of intermittent supply and many parameters that ought to be considered prior to starting work. The most basic approach, which should be used in poor IWS situations, is a visual leakage survey when the system is pressurised up to

12 hrs/day. This approach, along with the repairs undertaken, is the first step in the leakage management procedure.

The following example scenario demonstrates the choice of appropriate leak detection methods and the steps to be taken in a system that has some advanced control over the level of the IWS and is at level 4 in the IWS management process, with some basic planning of ALC activities. This method presented is derived from previous experience working in similar utility networks. The method assumes both day and quiet hours' operation.

Scenario 3.1 is an urban area of 1,500 service connections. It has: 5 km of mains length of mixed material; a majority of domestic properties and a small number of commercial properties; residential streets and main roads with an intermittent water supply regime; and operating pressure of below 10 m/15 psi or 1 bar in pressure.

Method steps:

- determine the volume of leakage in the area based on the number of hours of supply;
- obtain a GIS plot of the area, incorporating the boundary and any associated normally-closed valves, if applicable;
- check records from recent leakage management activity in past surveys.

ALC Step 1 (can be used in all IWS regimes)

- complete a sweep of the area with a visual leakage assessment only to find the simple and easy leaks where no experience or technology is required;
- investigate the visual survey for water running into drains or sewer channels.

ALC Step 2 (can be used in IWS of 4 hrs and above, and best after tanks filled)

- initial leak noise location undertaken by sounding all services using an electronic listening device when the system is pressurised, to locate 'areas of interest' (manual listening sticks should not be used as they are not as effective as electronic listening devices in poor pressure areas, plus leaks can be missed);
- use a simple 3 + 1 approach for leakage investigation:
  - first sweep: walk the area to view visual leaks and see any immediate anomalies;
  - second sweep: walk at time of maximum pressure listening on all meter and service connections, including any mains fittings such as valves and fire hydrants;
  - third sweep: walk the route of the main at time of maximum pressure with staff on both sides of the road listening every 1 m with a ground microphone;
    Should areas of interest not be located
  - pass back for a Step Test to be designed and completed and, after it is completed and results found, repeat sweeps 2–3
    Should areas of interest still not be located
  - Pass back to Leakage Supervisor or Project/DMA supervisor to hand back to necessary team to strip down to understand if there is an anomaly causing the high night flows, introduce smaller step tests or recalculate property count, allowance etc. and when a positive result is found pass back to leakage team for investigation.

ALC Step 3 (to be used in IWS above 8 hrs when tanks are being filled and the system pressurised)

- use a simple 3 + 1 approach for leakage
  - first sweep: walk the area to view visual leaks and see any immediate anomalies;

- second sweep: walk at time of maximum pressure listening on all meter and service connections, including any mains fittings such as valves and fire hydrants;
- third sweep: walk the route of the main at time of maximum pressure with staff on both sides of the road listening every 1 m with the ground microphone

  Should areas of interest not be located:
- pass back for a Step Test to be designed and completed and, after it is completed and results found, repeat sweeps 2–3

  Should areas of interest still not be located:
- pass back to Leakage Supervisor/Project/DMA supervisor to hand back to the necessary team to analyse to understand if there is an anomaly causing the high night flows, to introduce smaller Step Tests or recalculate property count, allowance etc. and, when a positive result is found, pass back to leakage team for investigation using either traditional ALC approach or some other more advanced technology.

3.5 ASSET MANAGEMENT

The operation of a water network under IWS increases the rate of asset deterioration and pipe bursts, leading to higher leakage rates. Water network topography and components are designed to operate under pressurised conditions. The sudden variation in flows and pressures under IWS, as well as the repeated dry and wet conditions, accelerate the deterioration of the pipe network as well as water meters. Additionally, detecting and repairing leaks becomes exceedingly challenging as low and changing pressures render leak detection and control almost impossible.

Although intermittent systems are generally designed to operate 24 × 7, they are forced to operate intermittently with different flow conditions. There is frequent wear and tear on valves, so more manpower is needed for network repairs. The quantity of water to be made available over 24 hours has to be made available in fewer hours in an intermittent system, which requires distribution pipes with larger diameters.

Once intermittent service becomes the norm, the hours of service continue to decline. The high costs of intermittent supply are paid by the utility, which incurs higher investment and operating costs – which ultimately the utility will not be able to meet. Furthermore, consumers have to pay the costs – so called coping costs – for additional facilities, such as storage tanks, pumps, alternative water supplies and household treatment facilities. Wealthy consumers cope by spending money on water tanks, pumping systems and filters, whereas middle-income groups spend less on capital equipment but more in terms of time and power. For the low-income group, however, the coping cost is primarily the opportunity cost of the time they must spend collecting water.

With the issues of customer metering and subsequent billing and payment as described above, water utilities are consequently often running at a loss and depend on subsidies from the government sector. This then leads to a lack of available budget, leading to the networks deteriorating further on a year on year basis, until they get so bad that outside funding is required to either renew or improve the existing assets.

3.6 WATER BALANCE AND KPIs

Ideally, a full water audit should be carried out, resulting in an IWA Water Balance with 95% confidence limits. But if time, funding, or expertise for a full water audit is not available, at least an initial audit, or 'Rapid NRW Assessment' is required.

A Rapid NRW Assessment is a simple, fast and usually inexpensive methodology to get a first understanding of the NRW situation in a water utility. It is a precondition for understanding the magnitude of the IWS's problems and the possibilities of transitioning to 24 × 7. A suggested procedure for the assessment is provided below:

- collect all necessary information required to determine the following:
  - daily volume of NRW;
  - average supply time (in case of intermittent supply);
  - average pressure;
  - number of service connections;
- determine the annual average daily volume of NRW. Review available records and identify possible irregularities, and determine the most likely average daily volume. Review billed consumption data (both billed metered and billed unmetered consumption) and determine the annual daily average volume of billed consumption;
- calculate the average daily supply time. Identify areas with different supply times, determine the approximate number of service connections for each area and calculate a weighted average supply time using the number of connections as the weighting factor;
- identify areas with different pressure characteristics, measure pressures and calculate average pressure for each area; determine the approximate number of service connections of each area and calculate a weighted average pressure using the number of connections as weighting factor – please note that only pressure during supply hours must be used in the calculation. (Record pressures with electronic pressure loggers for at least a 24-hour period at each measuring point; the average pressure in a certain point is the 24-hour average. If no pressure loggers are available, pressures during low-pressure periods (normally during peak demand time) and during high pressure periods (usually night time pressures) should be measured with pressure gauges);
- determine the number of service connections (SC), which is in many cases not equal to the number of customers (only in cases where each customer (account) is supplied by an individual service connection does the number of connections equals the number of customers). If, for example, every single housing unit in a multi-storey building is considered to be one customer (and therefore one account), then the number of service connections will be significantly less than the number of customers (one building may, for example, contain 100 customers but be supplied by only one service connection).

If the volume of NRW is not known, the following additional assessments can be made:

- if the SIV is not metered, short-term SIV measurements with portable flow meters are required. It is important to get at least crude data on the average daily SIV;
- it is extremely important to get an understanding of actual water consumption. (In some water utilities, especially in South Asia, consumption is not metered and consumers are either not billed at all, or billed on estimates or on the value of their property). There is no other way than by installing water meters, ideally high accuracy meters with data loggers, in about 100 truly randomly selected domestic customers. The customers must be metered for at least one week and average consumption calculated (either per customer/day and capita/day). The consumption of non-residential customers (commercial, institutional, industrial) must be assessed separately. Once this is done, the area's estimated total water consumption (whether billed or unbilled) can be calculated.

3.6.1 Water loss performance indicators

After completion of a water audit or Rapid NRW Assessment, real losses performance indicators need to be calculated. Real losses are expressed in litres per service connection per day per meter (head) average pressure, and adjusted for the average daily supply time, as:

l/conn./d/m (w.s.p.)

where l = litres, conn. = service connection, d = day, m = meter, and w.s.p. = when system pressurised, i.e. under 24/7 water supply.

A step by step calculation is presented below:

(1) determine daily volume of real losses; various different options are possible:
  - water audit available: simply take average daily volume of real losses from the water balance;
  - metered consumption data available but no water balance: assume volume of real losses = 90% of the volume of NRW;
  - Consumption data not available and consumption measurements have been carried out: real losses (RL) = SIV – calculated volume of water consumed

(2) divide average daily real losses (RL) by the number of service connections and express result in litres/connection/day (RL in l/conn./d)

(3) adjust to supply time:
  - PL l/conn./d (w.s.p.) = [RL l/conn./d]/[average supply time (h/d)] × 24 h

(4) divide by average pressure:
  - [Present RL l/conn./d/m (w.s.p.)] = [RL l/conn./d (w.s.p.)]/[average pressure (meter head)]

3.6.2 Volumetric assessment

This is one of the most important, and often overlooked, aspects in planning the transition from IWS to 24 × 7 supply.

Increasing the supply hours into a network under IWS would certainly result in an increase in leakage and possibly in an initial increase in consumption. The additional volume that would be needed is calculated as follows:

- determine how much additional water (system input) would be required to supply the area under 24 × 7 with the network still in its present, leaky, condition;
- set a target leakage level for the rehabilitated network;
- calculate the volume of real losses and then the total water demand for 24 × 7 supply for the rehabilitated network.

A step by step guide to each of these is presented below:

(1) Additional volume of water required:
  - Volume of real losses for present network condition under 24 × 7 supply:

$$\begin{aligned}&[\text{Volumeof real losses present (l/d) (for } 24 \times 7)]\\&\quad = [\text{Number of SC}] \times [\text{Present RL l/conn./d/m (w.s.p.)}]\\&\quad\quad \times [\text{expected average pressure (meter head)}]\end{aligned}$$

(**Note:** it needs to be expected that pressure will increase when water is supplied 24 × 7; make a reasonable assumption based on the local situation).

- Volume of additional water required (l/d)

  = [Volume of RL for 24 × 7]−[Volume of RL for IWS]

(2) The target leakage level will depend on the present leakage level. As a first suggestion, the following formula might be used to get an idea about a realistically achievable real loss target level:

$$\text{Target RL l/conn./d/m (w.s.p.)} = [\text{present RL l/conn./d/m (w.s.p.)}] \times a + b$$

where:

a = target reduction index (i.e., 0.05 is 95% reduction of current real losses)

b = additional margin of achievable real loss target (i.e., 8 is an additional 8 l/conn/d/m)

Once this (still high) level of real losses has been achieved, further real loss reduction programmes can be designed and implemented to reduce real losses further until the most economic level is reached.

(3) Calculation of the future SIV at 24 × 7 supply, target level of real losses and target pressure, is as follows:

- Volume of real losses for future network condition and 24 × 7 supply:

$$\begin{aligned}&[\text{Volume of future real losses (l/d) (for } 24 \times 7)]\\ &\quad = [\text{Number of SC}] \times [\text{target RL l/conn./d/m (w.s.p.)}]\\ &\qquad \times [\text{target average pressure (meter head)}]\end{aligned}$$

(**Note:** the target pressure will depend on the hydraulic situation and the service policy level of the water utility).

- [Future SIV] = [Future Volume of RL] + [Consumption]

3.7 HUMAN RESOURCES

In a Basic Network System, the water utility places minimal emphasis on staff training or education and has no allowance in the annual budget for such activities. Training employees is about teaching them how to perform a specific task or procedure. It is usually focused on short-term gains, enabling employees to become better at their current job.

Adequate staffing levels and provision of proper training and development should be ongoing efforts that are made in a water utility to improve employee performance and self-fulfilment through a variety of educational methods and programmes. An improvement in this key area would include basic training for some operational field activities – mostly on-the-job training to start with, then gradually moving to full training – and sustained and adequate staffing levels.

Training efforts need to focus on building up NRW knowledge and capacity in order to kick-start system improvements based on proper system management, collective and coherent work effort as well as results-based activities.

Coordination between the various water utility departments is crucial, e.g. technical and commercial departments, in order to have a common understanding of the way forward and about the importance of climbing to the next operational level in a structured and sustained manner.

3.8 MOVING TO THE NEXT LEVEL

3.8.1 Transitioning a system to 24 × 7 supply

Transitioning to 24 × 7 supply will almost always include a major NRW programme, which requires reduction in both real and apparent losses.

Roadmaps for transitioning from IWS to 24 × 7 are valuable tools for aspiring and guiding change. Roadmaps can be drawn to look at the wider scope of tackling water integrity at the sector level. Such roadmaps should focus on alternative and efficient water and energy resources, trunk main and storage capacity improvement, water use efficiency and consumer outreach and awareness, and tariff reform. Roadmaps can also be drawn to guide the detailed transition of distribution networks from IWS to continual service. Such roadmaps should focus on hydraulic optimisation, the application of network control and monitoring technologies, assessment of water balances during transition, customer metering, and network inspection and maintenance.

While transitioning from IWS to continuous supply through network rehabilitation and leakage reduction is the main current universally applicable conceptual approach, the evolution of demand patterns during and after transition poses a human factor that is not predictably quantifiable and is a critical challenge which remains largely unaddressed. User demand behaviour may differ depending on water prices, the provenance of unauthorised consumption, private storage, level of water conservation awareness, and many other local variables. Nuanced guidelines and tools for planning and implementing transition are required to help water service providers handle this and other complex aspects faced during the process.

Managing a water supply system that is intermittent and with high levels of NRW provides water utility managers with many problems. Although there is a generic approach to quantification and management of NRW based on IWA best practices, their application to intermittent water supply systems poses great challenges to practitioners.

Transitioning from IWS to 24 × 7 supply is different depending on the type of IWS system:

- was the system designed for IWS, meaning that it was always planned to bring water to consumers only once or twice per day? In these cases, networks normally have oversized pumping stations and trunk mains since the same volume of water needs to be supplied during a short period and therefore the flow rates during this period are high. In such systems, most customers (except the urban poor) have ground and/or roof tanks. As soon as the supply is on, all ground tanks get filled and therefore pressure in the pipes is very low.
- was IWS introduced recently in a system that was previously supplied on a continuous basis? This could typically be during an unexpected drought period when the utility's management was afraid that they could run out of water and hoped that, with IWS, water demand could be substantially reduced. In such cases, most customers don't have ground tanks and pressure will be normal during supply hours. But pressure peaks do occur when opening and closing to schedule supply to different areas (or to the entire system), causing extreme pressure fluctuations and pressure transients, which will have a massive impact on burst frequency.
- has the system always supplied 24 × 7 but some areas now only receive intermittent supply? This could be the situation where the intention is to supply the entire system 24 × 7 with continuously operated pumps but some areas which perhaps previously had continuous supply now receive water on an intermittent basis. Such systems, especially at points close to the IWS areas, are characterised by extremely low pressures during peak demand hours and high night-time pressures. Water usually only reaches the IWS areas during these night hours.

Transitioning to 24/7 will need different strategies depending on the different situations. The proposed methodology is suitable for the first and second situations outlined above. In the third case, a hydraulic analysis of the distribution network is required, so that hydraulic bottlenecks can be detected and areas with the highest levels of real losses (normally areas that are still supplied 24/7) can be identified. Reduction of real losses has to begin in these areas and water saved can then be supplied to other areas which at present experience IWS.

3.8.2 Planning the transition from IWS to 24 × 7

The last step is to decide whether there is enough water available to supply. Based on the volumetric assessment carried out as shown above, it can now be determined whether there is enough water available to immediately supply the entire system. In general, this will not be the case – otherwise, there would be no reason for IWS. Therefore, small isolated zones (DMAs) need to be established, over an area in the network where 24 × 7 supply is possible while the rest of the distribution network can still be operated intermittently. This will normally be close to a reservoir, treatment plant or along a transmission main which is always pressurised. Once real losses have been reduced, the water saved can be used to supply the next isolated zone (DMA), and so on.

Transitioning from IWS to 24 × 7 supply is not an easy task. It requires commitment and dedication from all concerned, including governments, water service providers and consumers. The following areas need to be addressed simultaneously in order to have a successful and sustainable impact:

- technical – gradual increase in the hours of supply aiming for continuous service, introduction of customer metering policies, improved network operation using DMA/sectorisation practices and targeted rehabilitation/replacement of mains;
- financial – implementation of tariff structures linked to performance incentives for saving water, cost recovery, adoption of commercial thinking and reform of water service providers to make them accountable;
- institutional – water service providers that have fallen into IWS have major governance and incentive flaws and need in-depth reform; moving to continuous supply often requires very difficult political and institutional choices that many governments prove reluctant to make. A paradigm shift is imperative!;
- social – water service providers need to gain the trust of their consumers, have the willingness to change, and involve the public in this effort in order to provide the required level of service at all times in a reliable and sustainable manner;
- communication with the customers – communicating all the above in an effective and convincing manner to all involved is of the utmost importance in order to have the maximum possible impact.

Transitioning from IWS to 24 × 7 supply is possible by applying the appropriate approach, techniques, methodologies and practices. However, it requires commitment and dedication from all concerned – governments, regulators, water service providers and, above all, consumers who are the recipients of the service.

3.8.3 Consumers and transitioning to 24 × 7 supply

Transitioning to 24 × 7 supply would certainly mean an improved level of service for consumers. However, it is of the utmost importance that consumers be kept informed from the very beginning, that they be made aware of the transition process well in advance, and that their support and cooperation is enlisted. Consumers may find it very hard to support transition unless they trust that the water utility's proposed actions will provide a better and sustainable service. It is important to communicate with the consumers at all stages to alleviate concerns about many key issues that may arise during transition, such as:

- which areas will get 24 × 7 supply first? Will it be introduced area by area, one area after the other? If not managed correctly – both technically and with good communication – customers in areas not yet receiving 24 × 7 supply may get impatient and angry. Tariff structures may need to be reviewed and increased accordingly, which of course needs to be communicated to consumers from the very beginning to make them aware that the improved service that they will receive will have an additional cost;
- will all households in an area, have the existing plumbing to receive 24 × 7 supply? The water utility should check service connections to all premises ensuring that all users are supplied with a single registered connection and properly metered. In addition, the utility should assist consumers to amend or adjust their existing plumbing connections to suit the continuous supply regime, thus avoiding wastage;
- will water quality be better or worse? Consumers need to understand that water quality with 24 × 7 supply should be better, but there may be an initial period of increased leaks and pipe bursts, possibly causing quality and supply problems. There may be some discoloration of water due to leakage repairs. Concerns can be alleviated easily through a good communication programme.

Consumers may use more water when 24 × 7 supply is established. This may be just for a short period of time until they adjust to the new supply – which can cause problems for the utility in supplying enough water but also cause concern to consumers themselves, due to the higher bills they may incur, particularly in cases where there is a block tariff in place. In many instances, low income consumers do not receive enough water to meet their needs under IWS, so an increase in water use with 24 × 7 supply would be expected and would in fact be desirable to improve public health.

Therefore, it is prudent to have a functional demand management programme in place in order to harness demand within reasonable levels and to sustain continuous supply, bearing in mind that any water conservation programmes should consider those who have always had very little water. This is just one example of how it is extremely important to segment customers when planning to move to a 24 × 7 supply. Some customers will actually want to use more water, but not too much more due to costs, and awareness campaigns to help them continue to use the water wisely will be needed. Programmes for household water conservation at a wider level will need to be carefully targeted to higher consumption customers who may already be wasting water and do not have cost constraints. These customers are the ones most likely to use much more water when there is a 24 × 7 supply.

Promising 24 × 7 supply and providing it for a time before reverting back to IWS is a huge problem. Raising customer expectations is never a good thing for any business. In addition, customers may have removed their water storage tanks, and new households would not have storage systems at all. This unfortunately would greatly reduce trust in the water service provider and cause undue harm to customers.

Many studies worldwide have shown that consumers are willing to pay for services provide they receive proper and sustainable service. Politicians, however, can sometimes be averse to changing the status quo and often provide greater resistance to moving to 24 × 7 than the utility and the consumers because they do not want to be unpopular.

There may also be other people or groups with vested interests in keeping IWS. Private vendors sell water to consumers who do not receive enough water from their utility. These perfectly legal water vendors who have come in to provide a service where the utility fails to do so will lose business with a

24×7 system. Private vendors are a very strong lobby and can resist the change to 24×7 in order not to lose a lucrative business.

3.9 DANGERS OF REVERTING TO IWS

Maintaining continuous supply after transition from IWS depends on optimising the operation of the network and minimising water losses. Continuous supply creates an additional burden for Operation and Maintenance and an initial increase in NRW. The utility must rise to the challenge and put resources and budgets in place to deal with the situation, even mobilising external financial assistance to achieve continuous supply, if needed.

The challenge of transitioning to 24×7 supply area by area may reveal information gaps that can exist in a water network and may prove that achieving full DMA operability across the entire network represents a significant challenge to the utility. The very act of isolating a DMA will result in enhanced knowledge of the network connectivity. With persistence, this can be replicated across the entire network and will then place the utility in a better position to manage their network and sustain continuous supply.

Upgrading and replacing water meters is also important. If the existing consumer water meters are not functioning properly and 24×7 supply is introduced, there will be billing errors resulting in lower revenues for the water utility.

There is a very strong case for greater consumer engagement through strong communication prior to continuous supply implementation. Consumers need to understand that with increased water availability and usage comes higher bills. Through carefully managing their usage and conserving water the increased number of consumer complaints could be reduced.

Usually, moving from IWS to 24×7 is part of a comprehensive programme to reduce NRW, with strong emphasis on reducing and controlling real losses but also including apparent losses to address consumer metering and billing issues.

While it is relatively easy to turn a 24×7 system to an intermittent supply, it is very hard to do the opposite. Water utilities that have fallen into the vicious cycle of IWS have major institutional, technical and financial issues and would definitely need to go through a reform process; moving to continuous supply often requires very difficult political and institutional choices that many water utilities/ governments prove reluctant to make. A paradigm shift is therefore imperative as part of a successful transition from IWS to 24×7 supply.

Water utilities in many countries around the world running their networks on intermittent supply are in need of assistance to improve their services. The key to better service is generally enhanced and improved management of systems, not just more capital expenditure on assets. Yet for the vast majority of these utilities, there is no clear route to management reform. Traditional training and capacity-building approaches are indeed necessary to build a solid foundation on which a utility can achieve long term sustainability and improved levels of service, while Public–Private Partnerships could provide the vehicle for transitioning from IWS to 24×7 supply in a cost efficient and effective manner. In many countries, this approach may still not be feasible or politically acceptable but it has generally been gaining wide acceptability in recent years due to successful applications.

Chapter 4

Ordinary network systems

An ordinary network system is one that is normally pressurised continuously, with no periods of zero supply, ensuring that all customers receive a constant water supply. However, there may also be some cases of otherwise intermittent network systems that are more developed in some key areas – with the result that they are in fact classified as an ordinary network system.

4.1 BULK FLOW MANAGEMENT

Installation of system input meters is essential for a water utility to understand how much water is being fed into the supply system. This measured amount of input volume can then be used to understand system demands and more importantly undertake a more accurate water balance to understand current NRW levels.

With a continuous water supply, the installation of system input meters is an easier decision to make than it is in the case of an intermittent system, as the continuously pressurised pipelines ensure the continuous accuracy of the flow meters.

With most system input meters being located at or close to a WTP, it is quite easy for WTP operators to take a manual reading on a regular basis. But it is also relatively easy to hard wire a connection that feeds data from a meter transmitter to the WTP Supervisory Control and Data Acquisition (SCADA) system. This then enables automatic monitoring of SIVs to be recorded at frequent intervals.

A next step of automation would be to have individual WTP SCADA systems sending regular updated system volume data to a centralised control centre, which can then start to monitor the supply system as an entirety.

4.2 CUSTOMER METERING

Intermittent water flow in supply and distribution networks may result in the deterioration of water meter accuracy, ultimately leading to malfunction due to high water velocities developed during the filling of

Author(s): Stuart Hamilton, Bambos Charalambous and Gary Wyeth
doi: 10.2166/9781780409207_0041

pipelines, to the transport of suspended solids in the system, and to the expulsion of air through the meters which can seriously damage their registration mechanism. This situation affects most of the water meters installed in supply networks whether they are used to measure bulk water flows or customer consumption.

Under intermittent supply, water utilities may neglect customer meters and associated activities due to the many other serious problems caused by such an operational regime. Moving from intermittent to continuous supply, a water utility may be in a situation where not all customers have meters installed and the utility has no regular replacement policy in place, replacing only stopped or malfunctioning meters. Meter reading activities are not properly checked and there is no system controlling meter readers themselves.

As water supply conditions are improved under a continuous system operation, the customer database should be periodically updated and an illegal connections programme initiated, with periodic checks for fraudulent activities. Moving up the customer metering improvement ladder, a meter replacement programme should be put in place, with the customer database regularly updated and a thorough illegal connections detection programme in place.

Reaching a level where all customers are metered by accurate meters, together with a proper replacement programme and meter reading system in place (consisting of handheld devices linked to the customer database, with alarms and warnings for abnormal readings) signifies a sound and robust situation, indicating a higher level of operations and system intelligence.

4.3 PRESSURE MANAGEMENT

Once a water utility moves to an ordinary network system, the main issue faced is maintaining the continuity of supply. Other issues, such as metering and real loss control are often not too important and take a back seat. In addition, if a water supply system has previously been operating in an intermittent state, the condition of the pipes will be quite poor, with consistent variations in pressure having had a fatigue effect on them. Therefore, leakage will be a very prevalent problem but is something that the utility perhaps does not have the time, resources and budget to manage efficiently through ALC and repair.

One easy way of controlling leakage volumes is through pressure management, in which reducing system pressures effectively reduces the volume of water being forced out through the leak holes. It should be noted that this method does not find and repair leaks but does have the effect of reducing the frequency of new leaks and bursts occurring.

There are a few methods of managing system pressures but, most commonly, basic water utilities use their existing gate/isolation valves to throttle pressures down in certain areas. However, this is probably the worst way to control pressures, as it mainly reduces pressure during peak demand and has no real effect on pressure during low demand periods. In addition, often these ‘throttled’ gate valves are forgotten about or even buried, meaning that they remain throttled continuously and thus system pressures remain artificially reduced.

More effective methods of reducing pressure use Pressure Reducing Valves (PRVs) or, in the case of pumping systems, use a variable speed control pump system. A basic PRV can reduce upstream pressure to a fixed downstream setting, the level of which needs to be determined by the water utility. This is often done by analysing the pressure at the minimum pressure point, in the zone being fed, and calculating what pressure set at the PRV gives sufficient pressure at the minimum pressure point during peak demand.

This method of controlling pressures using a PRV obviously only works if the PRV feeds a closed zone and higher pressures cannot feed through another pipe, which would negate the effect of the PRV. Many utilities implement Pressure Management Zones (PMZs), which are isolated zones with controlled pressures within them. Each individual PMZ can have its pressure managed independently, which depends on a number of factors including length and size of mains, number of connections, peak demand, topography, minimum pressure requirement and available inlet pressure.

Some PMZs may require more than one PRV to supply it, due to their size or peak demand requirements. When you have a PMZ with two or more PRVs feeding into it, then the setting of these PRVs becomes more critical to ensure that they don't start to 'fight' or 'hunt' each other (when the outlet pressure of one PRV starts to oscillate as it competes with the pressure of a second or third PRV).

Another commonly used method of controlling system pressures is to install pumps with Variable Frequency Drives (VFDs) or Variable Speed Drives (VSDs). These pumps are turned using motors, the speed or frequency of which can be adjusted to increase or decrease the outlet pressure of the pump. This type of pump control is often used in systems where direct pumping to customers is in place. Initially, these pumps are controlled on a simple time basis, either automatically or manually by the operator, and outlet pressure is reduced at night when demand is low and increased in the daytime when it is high.

4.4 LEAKAGE MANAGEMENT

Conducting leakage management and participating in ALC practices is considerably easier when in a 24/7 system; however, if the pressure in the distribution system is lower than 10 m/15 psi or 1 bar in pressure, then conducting ALC practices using acoustic methods can be a problem.

Conducting ALC activities when the pressure is at the highest and when the background noise is at the lowest is always preferable. The pipe material and pipe diameter also need to be considered when completing any ALC acoustic survey, as these will have an effect on the survey's effectiveness.

The following two scenarios demonstrate the choice of appropriate leak detection methods and the steps to be taken in a system that has a 24/7 supply regime to move it from the basic Level 1 to the more advanced Level 3 within the Water System Improvement Matrix. The methods presented are derived from previous experience working in similar utility networks, and assume both day and quiet hours' operation.

Scenario 4.1 is an urban area of 1500 properties. It has: 10 km of mains length predominantly of metallic construction; a majority of domestic properties and a small number of commercial properties; residential streets and main roads with a 24-hour water supply regime; and operating pressures above 10 m/15 psi or 1 bar in pressure.

Method steps:

- ascertain the current level of leakage from existing area metering and provide data for entry and exit leakage levels, when to start and stop ALC activities;
- obtain a GIS plot of the area, incorporating the boundary and associated normally-closed valves;
- compile recent leakage management activity from past surveys;
- carry out initial leak noise location through sounding and correlation techniques, to locate the 'area of interest';
- pinpoint the leak location and issue the ticket to repair;
- re-evaluate the level of leakage, taking into consideration the estimated losses from the number of leaks located. If this is not sufficient, continue surveying the region until the estimated number of Equivalent Service Pipe Bursts (ESPBs) is located and the exit level has been achieved.

If no more leaks can be found using acoustic methods, then use the following procedures:

- locate and sound (listen to) all fittings within the zone;
- compile a 'step test' strategy and associated 'risk mitigation' plan for client approval;

- carry out the 'step test' strategy in accordance with the approved 'risk mitigation' plan;
- locate areas of high losses and concentrate leakage location and repair activity into these areas;
- continue with the above strategy until exit levels have been achieved.

Note: Because of the potentially inherent water quality risks associated with carrying out a valving operation within a mainly metallic-based densely populated network, 'step testing' should only be proposed after all other methods to achieve the exit level have failed.

Scenario 4.2 is a rural area of 200 properties. It has: 30 km of mains length of a mixed mains material; isolated properties and troughs; a 24-hour water supply regime; and operating pressures above 10 m/15 psi or 1 bar in pressure.

Method steps:

- ascertain the current level of leakage from existing area metering and provide data for entry and exit;
- obtain a GIS plot of the area incorporating the boundary and associated normally-closed valves;
- compile recent leakage management activity from past surveys;
- compile a 'step test' strategy and associated 'risk mitigation' plan for client approval;
- carry out the 'step test' strategy in accordance with the approved 'risk mitigation' plan;
- locate areas of high losses and concentrate leakage location and repair activity into these areas;
- carry out leak noise location through sounding and correlation techniques in 'areas of interest';
- pinpoint the leak location, and then issue a ticket to repair;
- re-evaluate the level of leakage, taking into consideration the estimated losses from the number of leaks located. If this is not sufficient to account for the leakage volume, then continue surveying the region until the estimated number of ESPBs is located and the exit level has been achieved;
- if no more leaks can be found using acoustic means, revert back to another step test;
- continue with the above strategy until exit levels have been achieved.

Note: Should the water supply be intermittent, ensure that the ALC work is completed after a period when water is in the supply and that the time period is sufficient for all tanks to be filled so that this activity will not be misconstrued as leakage.

4.5 ASSET MANAGEMENT

A water utility's moving to continuous supply is normally dependent on significant funding to enable new assets to be installed or existing assets improved, to ensure that the continuous supply will be effective and not have an adverse effect on NRW levels.

Part of these asset improvements will be the installation of a series of customer meters, to ensure that full revenue is being recovered for water used. This will help the water utility to understand what revenue it has to spend on future asset improvements and on NRW management activities. Initially these budgets will be aimed at improving the critical asset improvement programmes, such as meter replacement, valve repair or replacement and leak repair.

The water utility can also start to look at long term asset improvement plans, aimed at repairing or replacing more expensive items such as pumps, reservoirs and mains. These plans can be spread out over a number of years, dependent on yearly budgets available or, if seen as urgent, they could be bought forward by utilising funding options.

To understand fully its assets and their condition, the water utility needs to start maintaining detailed assets records. Initially, these may be in the form of paper records but with the ease of digital GIS packages it is now easy to convert all of these paper records into a digital GIS system. This GIS system should not only detail the exact location of each asset, but have linked to it a database of asset information, such as date of installation, material size, etc. This GIS system can also be linked to a leakage database, indicating where all leaks have been found and repaired. This can help to identify hotspots of leakage activity and thus issues in pipelines or in transient activities.

Once a fully digital GIS system has been produced, it can be used to fine tune the long-term asset management plan, identifying which assets are at high risk of failure and their criticality to the supply system.

4.6 WATER BALANCE AND KPIS

It cannot be stressed enough that water utilities must target their actions and investments in order to get maximum benefit. To achieve this, it is important to have the necessary expertise either internally or externally in order to be in a position to justify a proposed improvement plan, which above all should be financially viable and sustainable. Obviously, the right level of knowledge and experience are also required to carry out such a plan.

Tackling whatever gives the quickest revenue return first is advised, since it will provide finance for longer-term savings. Always think of the long-term returns and do not get caught up immediately in expensive solutions because they seem more attractive.

It is important to understand that a Water Balance is the starting point for any NRW work. It provides sufficient information to assist in the drafting of an improvement plan in order to move ahead with water loss reduction and, in parallel, to make strategic improvements to a network. In the past, particularly in developing countries, it was thought necessary to first develop DMAs in order to drive a NRW reduction master plan. Rather, we suggest that this could be done in parallel and that, in the initial stages, a Water Balance is in fact the vehicle for driving a NRW Master Plan.

The main findings from a Water Balance should:

- assist in estimating the best return with the minimum of investment in the shortest time possible;
- form the basis for planning NRW reduction activities;
- provide sufficient information for an effective action plan.

A Water Balance is a useful tool which, if used correctly, will certainly point the way to the actions and measures that need to be taken to reduce NRW. Answers to the following questions will take you to the next stage – from desktop to the field environment:

- what do you do with a Water Balance after the various components have been calculated?
- how can you use the numbers to work out an investment strategy and an action plan?
- how do you prioritise your actions in order to get the best return on your investment?

Most water utilities use a Water Balance to calculate NRW and to find the amount of water being lost. It is obvious that this is extremely useful, and must be assessed in order to have a clear picture and to account for each constituent component of the Water Balance.

The planning of NRW reduction activities to be carried out and ultimately the compilation of an action plan are based on the findings of the Water Balance and in particular on its main components, namely: authorised billed consumption, unbilled authorised consumption, apparent losses and real losses. Depending on the amount of water which is being lost in each of these components which comprise the

Non-Revenue Water, the action plan can be targeted in order to provide the best return with the minimum of investment in the shortest time possible.

Accounting of water is extremely important in this process and is achieved through a validated Water Audit which could be carried out internally by experienced water utility personnel or by an external auditor. A water audit is a critical first step in the establishment of an effective water loss management programme. With the successful completion of a system water audit, the utility gains a quantified understanding of the integrity of its distribution system and can begin to formulate an economically sound plan to address losses. Water loss in a public water system can be a major operational issue. NRW components can significantly affect the financial stability of the utility. Addressing the issues associated with the non-revenue components will certainly entail a significant cost for the utility.

The economic trade-offs between the value of lost water, given that it generates no revenue, and the investment to reduce this loss requires careful planning and economic judgment. The utility needs to clearly understand the type of loss as well as its magnitude. Water resources, and financial and operational consequences, must all be weighed when considering these issues, and the decision taken is unique to every system.

The Performance Indicators recommended by the IWA for NRW and which are derived from the Water Balance calculations are shown in Table 4.1. They are each described in more detail in the sections below.

4.6.1 Apparent losses expressed as a percentage of authorised consumption

Apparent losses are expressed as a percentage of the authorised consumption, representing the volume of water which was used for which no revenue was received by the water utility. To minimise this percentage, accurate customer meters need to be employed together with appropriate data handling. Thorough inspections are required to be carried out in order to immediately replace all stopped meters, install meters on any unmetered connections and minimise the illegal connections. Accuracy tests on existing water meters need to be conducted in order to ascertain the current meter under-registration and take relevant measures to minimise or eliminate this. Meter readings must be checked for errors by the billing department.

Table 4.1 IWA Performance Indicators.

Component	Type	Performance Indicator
Apparent losses	Operational	Percentage of authorised consumption
Real losses	Operational	Real losses as litres/service connection/day when the system is pressurised
Real losses	Operational	Infrastructure Leakage Index (ILI)
Non-revenue water	Financial	Volume of NRW as percentage of system input volume
Non-revenue water	Financial	Value of NRW as percentage of annual cost of running system
Non-revenue water	Operational	Litres/service connection/day when the system is pressurised

4.6.2 Real losses in litres per service connection per day or m^3/km of mains/day (w.s.p.)

This performance indicator is recommended for assessing the effectiveness of operational management of real losses in distribution systems. The choice for this performance indicator of either 'litres/service connection/day' or 'm^3/km of mains/day' depends on the density of service connections. If it exceeds 20 service connections per kilometre of mains, component analysis for well-managed systems has shown that more than half the volume of real losses is generated from leaks associated with service connections, and 'per service connection' is therefore the preferred indicator. If the density of service connections is less than 20 per km of mains, it is preferable to express real losses in 'm^3/km of mains'.

4.6.3 Infrastructure Leakage Index

The Infrastructure Leakage Index (ILI) is the most appropriate performance indicator for comparing performance in operational management of real losses. It is a non-dimensional index (current annual real losses/unavoidable annual real losses) which assesses the overall efficiency of management of real losses in the system infrastructure (up to customer meters) at the current operating pressure.

4.6.4 NRW by volume (%)

Although this PI is widely quoted, it is not necessarily particularly meaningful as it is strongly influenced by consumption. It is expressed as a percentage of the SIV that does not yield any revenue for the utility.

4.6.5 NRW by value

Because of the problems of interpreting NRW as a percentage by volume, it is recommended that NRW should also be expressed as a percentage by value – of the cost of running the system. Because the different components of NRW (unbilled authorised consumption, apparent losses and real losses) can have widely different values per unit volume, NRW percentage by value may differ significantly from NRW percentage by volume. It is usual to value apparent losses based on the sale price of water to customers. The lowest valuation for real losses can be based on the variable production and distribution cost (or bulk purchase price plus distribution costs), plus other costs as appropriate (deferred capital costs, environmental costs), and unbilled authorised consumption can be valued somewhere between the valuations for apparent and real losses.

4.6.6 NRW in litres/service connection/day (w.s.p.)

This Performance Indicator is recommended for assessing the NRW in terms of litres/service connection/day.

4.7 HUMAN RESOURCES

It must be appreciated that in order to have an efficient and effective system operation, appropriate levels of properly trained and supported staff must be in place. The required number of staff as well as their level of education, knowledge and skills needs to be established.

It is important that clear lines of authority are established to enable smooth and robust system operation. Under a continuous supply regime, NRW-related activities need to be executed by a dedicated team or department properly structured, staffed and supported to handle all matters relating to NRW activities.

Restructuring is usually inevitable and it is best to take place as early as possible. Normally there will need to be a transition from the current level of operational activities to a higher level which should be carried out in a smooth manner, assigning existing staff with relevant knowledge capacity and filling any vacant positions by re-skilling other suitable utility staff or employing new staff.

Specialised and dedicated ongoing training of the different levels of staff is an integral part of the whole staffing process. Concepts such as target setting and accountability of staff through proper structuring are worthwhile introducing.

4.8 MOVING TO THE NEXT LEVEL

By the time a water utility has reached the last step of the Ordinary Network System ladder, it should have a number of electronic monitoring, measuring and recording solutions in place. However, these solutions are often controlled manually and any analysis is done in individual departments and requires significant human input.

As the water utility moves to become a Smart Network System, it will need to develop better SCADA/communication platforms which enable automatic data collection and better sharing of information through a centralised data centre, as well as better automatic analysis which will feed back into the supply system, enabling better control.

In order to jumpstart a Smart Network System initiative, the following steps are recommended:

- within the first few weeks: bring in knowledgeable individuals who can perform an objective assessment. Identify those areas within the utility in the most need of efficiency improvements;
- in the following 6 months: begin to plan a roadmap. Identify cases where low up-front investment can produce positive results over a relatively short period of time. These serve as effective initial pilot projects;
- in the following year: identify areas where Smart Water Network benefits can be expanded, and agree on the scope, budget and resources to be used;
- in the following 2 years: create a long-term sustainability plan for the programme, create succession, and deploy long-term monitoring and measurement.

Moving from an Ordinary to a Smart Network System is not something that can be rushed into or done quickly. It is important to understand the water utility's unique requirements and constraints. Quite often what works in one utility will not necessarily be successful in another.

Ensuring great technical support, either inhouse or via a third party, is essential to the success of any Smart programmes and should be a key factor in deciding which solutions to implement.

Chapter 5

Smart network systems

Water utilities that are moving into smart network systems are effectively enhancing their supply system through the implementation of information technology solutions. Data-gathering sensors are installed across the network to monitor, record and communicate data back to a series of software solutions for further analysis.

Many continuous water supply systems may have sensors installed across their networks, but often the data is acquired manually and analysis of this data is undertaken at a minimum, often within individual departments.

Moving into a smart network system, these sensors start to communicate automatically, through wired or wireless communication systems, sending regular data back to a control centre. This data can then be analysed centrally, enabling results to be distributed to relevant departments, assisting in improving their operational efficiency.

At the high end of a smart water system's functionality, the results of the data analysis can also be used to automatically control the supply system's operations. These automated controls still need human oversight but are the start of an intelligent control system.

5.1 BULK FLOW MANAGEMENT

As a water utility moves into being a smart system, it should have all of its system input meters installed, with a monitoring system that feeds regular data to a centralised control room.

As with most electronic equipment, the system input meters have a definitive life span and will start to lose accuracy and eventually stop working completely. Therefore, smart water utilities need to put a meter health monitoring system into place with a preventative maintenance programme.

Analysis of regular SIV data should be undertaken to analyse possible changes in meter accuracy or failures in system electronics. Any anomalies detected can be checked through a meter verification of the meter electronic system. Even without detecting any anomalies, this meter electronic verification should be undertaken at least once a year.

Author(s): Stuart Hamilton, Bambos Charalambous and Gary Wyeth
doi: 10.2166/9781780409207_0049

In addition, a smart water utility should start to undertake yearly volumetric testing of the system input meters. This can be done by utilising a second meter, such as a temporary insertion meter or a strap-on ultrasonic meter, or by comparing the flow through the meter to a reservoir volume rise or fall.

5.2 CUSTOMER METERING

The full potential of smart water metering goes far beyond the enablement of accurate meter measurement and billing. Three overall areas stand out: daily operations, asset management and end-user involvement.

Smart metering potentially also provides the ability to empower consumers by making monthly consumption data and other information available to them, with clear benefits to be reaped by utilities ready to embrace the digital transformation that smart metering brings and improve operations.

Smart metering can illuminate a distribution network, providing the knowledge needed to effectively target efforts and be proactive rather than reactive. Once smart meters are installed, the additional expense for a utility to collect daily or even hourly data is minimal compared to the added value offered to its entire value chain from the adoption of a more holistic digital approach.

Smart metering provides the basis for utilities to make fact-based decisions related to the daily operations surrounding their core tasks. The customer database is updated based on the information and data from smart meters and is linked to a GIS. Handheld devices for meter reading are used for meter reading and bills issued on the spot.

But the value of smart meter data goes far beyond accurate billing. The challenge so far has been to utilise its full potential but, with the right tools, it can do just that and transform meter data into actionable insights. The next step is to fully utilise an analytical platform that will enable the system to be used to provide options on 'how to' move forward, using 'what if' scenarios, where users can pick and choose what options to apply depending on their needs or interests. The meter data can be automatically organised, visualised and analysed to give a total overview of the distribution network and to provide the knowledge needed to effectively target efforts within leakage detection, water loss reduction and event management.

5.3 PRESSURE MANAGEMENT

Most smart water utilities have already undertaken basic pressure management practices and are now looking at how to improve their efficiency. They are generally now looking to develop Calm Networks, where maximum pressure and pressure fluctuations are minimised, ensuring minimal stress impact on assets.

System demands vary through a 24-hour period, from very low during the night-time period, to very high during the morning and evening peaks. Therefore, system pressures do not need to be constant across the entire 24-hour period, but can be adjusted as demands vary.

This modulation of the system pressures, either at PRVs or pumps, can be done in several different ways, namely time based, flow based and system pressure based.

In time-based modulation, the outlet pressure of a PRV or pump is adjusted based on fixed time settings. This can start with a very basic two steps, in which there is a low pressure at night and a high pressure during the day, to a more complicated system with several different pressures adjusted at multiple times of the day.

In the case of pumps, the VFD/VSD motors are adjusted to speed the pumps up or down, which can be done manually by an operator or automatically using the pump PLC control panel. As most pump stations have power, multiple adjustments are easily made.

For PRVs, time-based modulation is a bit more complicated and requires the addition of a controller to the existing basic PRV, with the capability of mechanically adjusting the PRV setting. Many PRV controllers are designed to operate using batteries, as most PRVs are installed in quite remote locations. Therefore, battery life and the effect on battery life with multiple pressure adjustments is something that must be carefully considered.

Flow-based pressure control involves adjusting the outlet pressure of the PRV or pump based on flow rate through the PRV/pump. This means that a flow meter needs to be installed directly after the pump or before the PRV to measure the flow. Once again, an electronic controller is required, connecting the flow meter and monitoring the through flow continuously. As flow changes, the outlet pressure of the pump or PRV can then be adjusted accordingly. For these controllers to work, there needs to be a relationship developed between flow and outlet pressure, so that the controller understands what adjustments to make as flow changes. This relationship can be manually developed and entered into the controller or can be automatically developed using smart algorithms.

In system pressure-based control, flow control is managed to achieve a certain set pressure at some point in the PMZ. Generally this is at the minimum pressure point within the PMZ, since if good pressure is supplied there then the entire PMZ will have good pressure. A permanent pressure sensor is installed at the control point to monitor system pressures which are fed back to the utility where the flow vs outlet pressure relationship can be further refined.

This development and refinement of the flow vs outlet pressure relationship can be done manually but becomes more intelligent if done automatically. Here the relevant flow and pressure data is uploaded to a central server that then refines the relationship automatically and sends revised instructions back down to the controller. In this way the system can modify itself as changes in demands occur on a daily basis, ensuring that optimal performance of the PRV or pumps is maintained.

5.4 LEAKAGE MANAGEMENT

The following scenarios demonstrate the choice of appropriate leak detection methods, and the steps to be taken in a system that has a 24/7 supply regime and has moved from the standard approach to a more advanced smart network. This method presented is derived from previous experience working in similar utility networks. The method assumes both day and quiet hours' operation.

Scenario 5.1 is an urban area of 1500 properties. It has: 10 km of mains length predominantly of metallic construction; a majority of domestic properties and a small number of commercial properties; residential streets and main roads with a 24-hour water supply regime; and operating pressures above 10 m/30 ft/15 psi or 1 bar in pressure.

Method steps:

- complete a DMA prioritisation procedure using software to understand the worst performing DMA and plan the ALC activities;
- understand the full level of the KPIs, along with the intervention and exit level, to ensure optimum management of the leakage team's performance;
- use permanent monitored leakage technology to look for areas of interest;
- proceed with office-based correlation to show the position of potential leaks;
- obtain a GIS plot of the area or interest to allow the leakage team to evaluate;
- pinpoint the leak location and issue a ticket to repair;
- re-run the prioritisation software and continue to drive down the leakage.

Scenario 5.2 is a rural area of 200 properties. It has: 30 km of mains length of a mixed mains material; isolated properties and troughs; a 24-hour water supply regime; and operating pressures above 10 m/30 ft/15 psi or 1 bar in pressure.

With high levels of established and permanent DMAs with permanent acoustic monitoring of systems, the procedure for either a rural or urban are does not change, since the robust KPI monitoring in place is used to prioritise any areas of interest to investigate. The methods used in Scenario 5.1 therefore apply equally in this scenario.

5.5 ASSET MANAGEMENT

As a water utility moves into being a smart network system, the management of its assets becomes a key component of its operational activities. Smart water utilities understand that if they can better manage their assets through improved maintenance, rehabilitation and replacement techniques, then the longer the assets will last and the lower the physical losses will be.

The key to managing these assets better is to understand the assets and their condition. Smart utilities will already have an accurate GIS mapping system which they can now start to develop, through the addition of historical leak and burst data, and with pipe data obtained from repair activities. Additional analysis such as asset risk assessment and asset condition/remaining life can also be undertaken to further enhance the picture of what condition the existing assets are in and their importance to the supply system as a whole.

All of this information can then be developed into a more accurate long-term asset management plan which the utility can follow to ensure steady improvements to its assets. This asset management plan should include regular asset maintenance requirements, regular asset replacement programmes such as for customer meters, and planned asset replacement such as of mains pipes.

A smart water utility should also start looking at various pipe rehabilitation techniques or 'no dig' technologies, to reduce the impact on customers and supply, and reduce traffic disruption, during pipe maintenance or replacement.

Whether undertaking leak repairs, mains rehabilitation or mains replacement, the principle aim should be to ensure a longer asset life and reduced physical losses over the long term. So, as well as developing a long-term asset management plan, a smart water utility needs to develop repair, rehabilitation and replacement methodologies that ensure this extended asset life. These should include selection of quality assets and materials, storage and transportation of the assets and proper installation following approved procedures.

5.6 WATER BALANCE AND KPIs?

The water balance and KPIs used in a smart network system are the same as those described in Chapter 4. However, the degree of human intervention is minimised in the case of a smart network system.

A water utility moving to being a smart network system will start to apply smart water management, adopting Information and Communication Technology (ICT) and real-time data and responses as an integral part of the solution to their water management challenges.

At utility level, the water balance will be updated regularly with latest billing and flow data, and used to prioritised NRW activities. At DMA level, the water balance will automatically be updated daily, utilising the automatic metering infrastructure in place to pull in customer consumption data, as well as DMA flow and pressure data. Relevant KPIs to benchmark and set improvement targets will be calculated automatically using relevant software linked to bulk flow measurements, billing and asset management databases. Water distribution management will be based on accurate, robust and up-to-date data and information, ensuring optimum use of water resources as well as proper allocation of water on an equitable basis to satisfy all uses and demands.

5.7 HUMAN RESOURCES

Having moved to the smart network level, a water utility needs to adjust its staff levels and training programmes to cater for the new technologies and systems in place. Staff training requires a holistic learning strategy to enable a culture of continuous learning and development, aligned with talent management. Deployment of a learning and development infrastructure leveraging learning management technology will be crucial in knowledge building, creating a robust and versatile work force properly responding to the challenges of an improved level of operation and performance.

Management and employees may have a developed knowledge of traditional water utility operations as well as legacy knowledge of ordinary systems. However, emerging smart technologies incorporating novel features can present new challenges for employees. They may lack familiarity with or simply misunderstand these new features and the extent to which they can improve utility operations and assist the utility in climbing further up the ladder.

It is important for management and employees to approach the adoption of new technologies with a mindset of improvement and development. To ensure thorough understanding, the utility should create workplace protocols in which capacity building, reskilling and knowledge transfer take place before the implementation of any new technologies, equipment and processes.

5.8 MOVING TO THE NEXT LEVEL

By the time a water utility reaches the last step of a smart network system within our improvement matrix, it should have a fully developed monitoring system, with multiple devices installed across the supply system monitoring for pressure transients and sound, and measuring flow and system pressures. The data from these devices should be fed automatically back to a central data centre, where multiple departments can view the data and use it to plan operational activities with better efficiency. There should also be the beginnings of some automatic data analysis, the results of which can be fed back, either manually or automatically, into the control functions of the operational department.

As the water utility moves to become an intelligent network system, they will need to develop a series of intelligent analytics, which will automatically analyse the collected system data and interpret the results to determine the most efficient operating control parameters. These control parameters will then be fed automatically back into the pumps, valves and other control devices to continuously enable efficient operation of the network.

The development of these analytics and the subsequent automatic control systems will require joint development with manufacturers to ensure the developed products meet the water utilities specifications and requirements. There would likely need to be financial investment at the development stage, as the manufacturers may be developing a solution that initially would only fit one or two water utilities.

Chapter 6

Intelligent network systems

Water utilities that are moving into intelligent network systems are those that are looking at developing more algorithmic analysis which is undertaken automatically with no or minimal human input. In addition, the output from these analyses is automatically used to change the water supply system control mechanisms, without human input.

Therefore, with the onset of these intelligent water network systems, we will have supply systems that are self-monitoring, self-analysing and self-adjusting the supply system to optimise performance, based on pre-determined parameters.

It should be noted that many if not all of the intelligent systems described in this chapter have not been developed as yet and are what manufacturers will aim to be developing in the hopefully not too distant future.

6.1 BULK FLOW MANAGEMENT

Intelligent water utilities would have accurate meters on all bulk flows into their network. These meters would automatically record data and send it continuously back to a central control centre. The flow data would be continuously analysed to ensure the accuracy of the flow meter, ensuring no drift or failure of the meter itself. Some utilities install two meters in series to ensure continuous flow monitoring if one of the meters was to fail. New meters will hopefully be able to re-calibrate themselves, with input from an intelligent control system.

It is also expected that an intelligent control system will be able to forecast demands, based on historical trends and current conditions. These forecasts would be used to control the flow into the system through the bulk flow meters, to ensure no oversupply.

Author(s): Stuart Hamilton, Bambos Charalambous and Gary Wyeth
doi: 10.2166/9781780409207_0055

6.2 CUSTOMER METERING

Conventional meters and traditional billing systems do not have the ability to cater for the unique requirements of users, but this could be possible in an intelligent metering system. A system where all customers are metered with high accuracy automatic meter reading (AMR) and advanced metering infrastructure (AMI) meters that have sound sensors to detect leak noises, and with connected communication systems allowing for automatic correlation and pinpointing of leaks, is a target for the future. Fraudulent activities could be detected via the AMI and AMR systems and automatic alerts sent to customers when excessive customer-side usage is experienced.

The ultimate goal is to employ complete machine learning algorithms that control the system, so that the system is run with internal flow measurement, measuring and adjusting meters for accuracy, with a total self-billing and bill collection procedure without need of any human interventions.

Future intelligent water metering systems will incorporate a number of key features that will make them attractive, such as the easy ability to upgrade the functionality of an installed meter base without removing the meter (reducing life cycle capital costs), a robust and secure meter management system to ensure 100% collection, two-way on-line communication with individual meters to change operating parameters with full programming flexibility, and the ability to accommodate unique local requirements such as multiple step tariffs, debt terms and payments.

From a customer perspective, an intelligent customer metering system will offer the ability to personalise the meter and allow self-management, offering a number of options for purchasing water or settling outstanding debt, as well as giving access to a full range of rebates and allowances offered by the water utility.

6.3 PRESSURE MANAGEMENT

Intelligent network systems have multiple pressure sensors installed across each PMZ, enabling monitoring and analysing of the pressures within the entire zone and not just at the critical point. The analysis of data from these pressure sensors is used to further refine the control algorithm for the pumps and PRVs. These control algorithms are used to manage the pump and PRV outlet pressures to ensure calm stable networks.

The data from the pressure sensors can also be fed into leak volume and demand analysis models and ultimately used to minimise leakage volumes by controlling system pressures, whilst ensuring sufficient demand. This automated process would be done through the intelligent control system with minimal human input.

6.4 LEAKAGE MANAGEMENT

The following scenario examples demonstrate the choice of appropriate leak detection methods and the steps to be taken in a system that has a 24/7 supply regime and has moved from the smart approach to a more advanced intelligent network. The method presented reflects the widely held expectations of how the procedure will likely work in the coming years.

Scenario 6.1 is an urban area of 1500 properties. It has: 10 km of mains length predominantly of metallic construction; a majority of domestic properties and a small number of commercial properties; residential streets and main roads with a 24-hour water supply regime; and operating pressures above 10 m/30 ft/15 psi or 1 bar in pressure.

Method steps:

- complete a DMA prioritisation procedure using software to understand the worst performing DMA and plan the ALC activities;
- cross reference other management performance systems and complete a cost analysis of the worst performing DMAs based on parameters such as cost of water, manpower, chemicals and distance to location;
- run an auto leak location procedure and support the leak location findings by some other non-acoustic procedure;
- initiate survey by drone or live satellite real-time data to confirm leak position and to plan for any road closure if necessary;
- confirm leak position in the live operating system, which will then plan the best time for automatic mains closure to allow repairs to take place;
- decide if excavation or internal mains repair will take place;
- continue to monitor and identify potential leak positions.

Scenario 6.2 is a rural area of 200 properties. It has: 30 km of mains length of a mixed material; isolated properties and troughs; a 24-hour water supply regime; and operating pressures above 10 m/30 ft/15 psi or 1 bar in pressure.

Due to permanent and intelligent system monitoring, the procedure for either rural or urban does not change as robust KPI monitoring is used to prioritise the areas of interest to investigate. The methods used in Scenario 6.1 therefore apply equally in this scenario.

6.5 ASSET MANAGEMENT

Intelligent water utilities will have a long-term strategy for asset management. Permanent sensors installed around the supply system will enable an intelligent utility to continuously analyse the condition of the network and thus continuously refine the asset management plan.

The development of intelligent pipe materials that have sensors built into them and have the ability to self repair will assist intelligent utilities and reduce their impact on customers.

For conventional pipe materials and those future pipe materials that do not have a self-repairing functionality, the development of internal repair technologies is critical. These internal repair technologies, such as robots or self flowing devices, will have the ability to undertake pipe repairs without the need to excavate a pipe and whilst the system is still supplying water, minimising customer disruption.

The ultimate goal is for complete machine learning algorithms that control the system and undertake analysis of the network to identify asset weaknesses and potential leak locations, and that analyses causes of failure and, where possible, rectifies the cause whilst being linked into the asset management system to prioritise systematic pipe replacement, including the optimum choice of pipe material.

6.6 WATER BALANCE AND KPIs?

The water balance and KPIs used for an intelligent network system are the same as those described in Chapter 4. However, the degree of human intervention is extremely limited in this case. Intelligent systems empower utilities to make insight-driven decisions and to provide optimum, cost-effective solutions to the most critical and complex challenges that could be faced by water utilities.

The intelligent network system concept incorporates smart supervisory control, instrumentation, and information management systems that are linked to provide optimum water supply and use. At a utility level, this enables total optimisation of a stable supply of safe water that meets demand, maintenance and operating efficiency goals in the cycle from water resources to water use.

All data and information to develop a water balance and measure KPIs are pulled in and analysed automatically from respective databases, and improvement actions prioritised by the system based on a cost-benefit analysis. The water balance is calculated daily, complete with accurate KPIs with full financial costings and with any errors adjusted based on machine learning algorithms.

6.7 HUMAN RESOURCES

Intelligent systems offer the next generation of solutions, powered by computing and artificial intelligence. Continued professional development of staff to improve and build knowledge and capacity in intelligent systems is crucial.

An improved staff training programme should be in place addressing cutting edge technologies and systems, coupled with coordination with research institutions in areas of system improvement using artificial intelligence and systems.

It is important to match staff levels and capabilities to complement the intelligent systems adopted. Artificial intelligence emulates the judgment and behaviour of humans and the skillsets, knowledge and capacity of the staff need to match this high level of intelligence.

Chapter 7

ALC in low pressure systems with both metallic and non-metallic pipes

This chapter describes how to conduct Active Leakage Control (ALC) in areas where a system has a 24-hour water supply regime and operating pressures below 10 m/15 psi or 1 bar. This new concept, developed in 2020, covers the procedure to use in any low-pressure system, with either a metallic or non-metallic distribution system.

7.1 INTRODUCTION

After 20 years working in several low-pressure systems where it was assumed impossible to locate leaks using acoustic means, a simple approach and rule of thumb has been developed based on findings from recent projects in Asia, Middle East, Caribbean and Canada.

When listening using an electronic listening device, the simple rule of thumb is that noise from a leak in plastic and polyethylene pipes will transfer 1 m distance along the pipe for every 1 m of pressure up to approximately 15 m in distance, with considerable noise attenuation irrespective of how high the pressure is. Beyond 15 m in distance, this linear rule does not apply. The major reason for this is that the noise of a leak – or its 'energy' – is absorbed into the soft pipe wall of plastic and polyethylene pipes, thus making it more difficult to hear the leak the further you are from the leak source, when a listening device is attached to a network fitting. The diameter of the pipe also affects the distance the leak noise will travel, with a maximum pipe diameter, suggested in our experience, of 200 mm when conducting ALC activities. For diameters above this, leak noise acoustic transfer is greatly reduced, meaning traditional ALC activities are not effective.

Following this principle, the ALC approach needs to be adapted depending on the pressure in the main and the distance between service connections and the length of service pipe to the main. It has been proved in practice that ALC can be undertaken in low pressure areas and can be successful if the correct approach is adopted. Findings from the initiative and the results from completed field trials are presented below.

Author(s): Stuart Hamilton, Bambos Charalambous and Gary Wyeth
doi: 10.2166/9781780409207_0059

7.2 HUMAN HEARING

The average range of human hearing is generally from as low as 20 hz to as high as 20,000 hz; however, 2000 hz to 5000 hz is considered to be the most sensitive range. The rate of hearing deterioration, which begins at an early age, has a direct correlation with working environment and the amount of loud noise a person is exposed to during their lifespan. The deterioration process starts generally around the age of 8 years old and by the time an individual reaches the age of a young adult, the top range has dropped from 20,000 hz to 16,000 hz, dropping again to 12,000 hz by the age of 30.

By the time an individual is in their 50 s, they can have lost the ability to hear both low and high frequencies, meaning they may not be able to hear certain sounds, and this is evidenced by the ringtones and volumes that older generations use on their mobile phones. Some ringtones available cannot be heard by older listeners, whilst still clearly audible to a younger person.

Deterioration in hearing is generally less pronounced in women than in men, with women having a greater range of hearing from their 20 s through to older age.

7.3 LEAK NOISE IS PRESSURE RELIANT

Many papers have confirmed that leak noise is pressure reliant – something that is likely not a surprise to anyone reading this chapter. In reality, it doesn't matter if you are listening along a pipe or above the leak, it is the pressure forcing water out of a hole in a pipe which creates what can be considered 'energy' and it is the noise of this energy as water leaves the pipe that we listen for. So, if the pressure is 50 m or 75 psi, then the amount of energy created will be much greater than if the pressure was at 5 m or 7.5 psi. When listening to a pipe itself or to a fitting, it should be noted that the transmittal of this 'energy' in the form of leak noise along the pipe is also reliant upon the pipe material and diameter.

7.4 METHODS OF CONDUCTING WATER LEAK LOCATION

In reality, the methodology of locating water leaks has not changed much since water mains were first introduced, when visual leak detection was the common approach. In the early 1900s, a listening device (similar to a doctor using a stethoscope) was developed that enabled leakage engineers to hear and locate an underground leak. Records of this come mainly from the UK; however, it is thought that many other

Figure 7.1 Depiction of a divining rod in use in Britain during the late 18th Century, from a volume by Thomas Pennant (Source: Independent newspaper, 22 November 2017; courtesy of the National Library of Wales).

countries may have used something similar. Prior to having a listening device, the engineers' 'advanced' approach was to use a water divining rod or dousing rod (see Figure 7.1 which shows a late 18th Century illustration of the technique). This method was not only used to find a water source but to also find a water leak.

7.5 EVOLUTION OF THE ROLE OF THE LEAKAGE ENGINEER

The individuals conducting ALC procedures have gone by many names and titles over the years: in the early 1900s until the mid-1980s they were called waste inspectors, with the latest terms being leakage engineer or leak detection technician. The term 'waste inspector' was derived from the concept that they were finding wasted water.

There are many stories linked to the first full time waste inspector and one that seems most likely is that a water pump operator lost his arm in an accident and, on his return, was not able to conduct any normal work so hence was sent out to locate leaks with a wooden listening stick, since this could be completed one handed. Workers were also offered a 'promotion' to be a waste inspector when they were too old to continue digging holes to either repair leaks or lay new pipes, keeping them in employment to go and look for wastage in the water distribution system. This approach continued and was still common practice in water companies up to the late 1990s, and in many cases is still the practice today.

What was not considered was the threshold of the hearing ability of an employee to hear the leaks they were asked to listen for through the non-electronic devices that were used, commonly known as a manual listening stick or manual stethoscope. These devices were made of wood or metal and varied in type; however, they all relied on the ability of the person to hear the leak noise. Electronic listening devices were first developed and introduced in the 1970s, although they were classed as a luxury item and were not in wide circulation for use.

The ALC procedure adopted prior to the 1970s was to either listen to a pipe on the mains fittings to hear the leak noise or to listen at every house connection – the latter being the location of smaller and service pipe leaks, whilst the 'mains fittings only' approach was used more for the location of burst mains. This approach was the common approach and is still common; however, the characteristics of distribution systems have changed.

Prior to the 1970s, most water supply main pipes were metallic and leakage detection was considerably easier since leak noise travelled much longer distances (depending on pressure). However, with the change of pipe material from metallic to non-metallic, a major issue today is that leak noise travel is considerably shorter, so that, in a mains pipe of plastic or polyethylene material, many leaks can be missed when sounding with a manual listening stick.

7.6 PRACTICAL APPLICATION

This chapter presents a new theory about how leak noise travels through plastic and polyethylene pipes at different pressures and the revised approaches that should be used when conducting ALC surveys.

As mentioned above, the common and standard approach when conducting ALC surveys is to listen on either mains fittings only or every fitting available, including all house connections, and the procedure in many countries around the world is to use a manual listening stick made of either wood or metal. This procedure is rarely changed regardless of pipe material, pressure in the main, or the ability of the leakage engineer to hear leaks being investigated.

By contrast, the revised procedure presented in the examples below is to listen at every service and house connection and – when a leak is heard – to pinpoint the position of that leak by surface sounding. The distance from the leak position to the service or house connection is measured and plotted onto a graph. The noise transfer by material, pressure and equipment is investigated and the results plotted. This investigative technique allows a new approach to be developed and thus creates a very simple rule of thumb guide to follow for future successful leak detection surveys.

7.7 RESULTS FROM PILOT AREA WITH PLASTIC AND POLYETHYLENE PIPEWORK

Figures 7.2 and 7.3 show the distance away from a leak that the leak noise could be heard using either a manual or electronic listening stick whilst conducting leakage detection surveys on plastic and polyethylene pipework.

The line in Figure 7.2 represents average results using an electronic device. The ability to hear a leak in relationship to pressure and distance is nearly proportional until around the 15 m point and then starts to tail off. The maximum distance that a leak could be heard was approximately 28 m from the source of the leak to the service or house connection with the pressure being 45 m.

In Figure 7.3, showing results using a manual listening stick, the numbers are not at all proportional and can be considered quite scattered. The results are considerably poorer than when using an electronic device, and it was found that the maximum distance that a leak could be heard was approximately 16 m from the source of the leak to the service or house connection with the pressure being 35 m. The most probable reason for this is that the leak noise is very quiet, and the

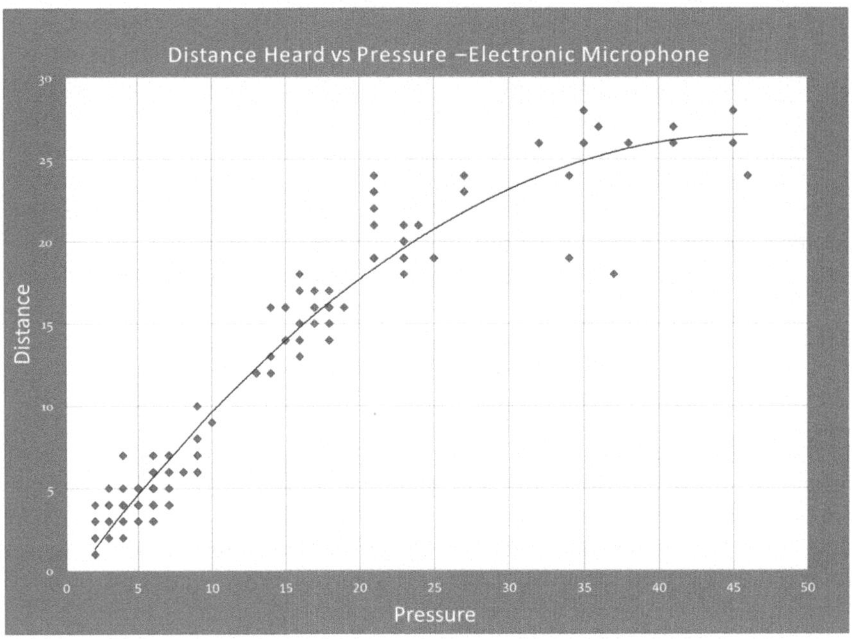

Figure 7.2 Plots of leaks heard using an electronic microphone (Source: Stuart Hamilton).

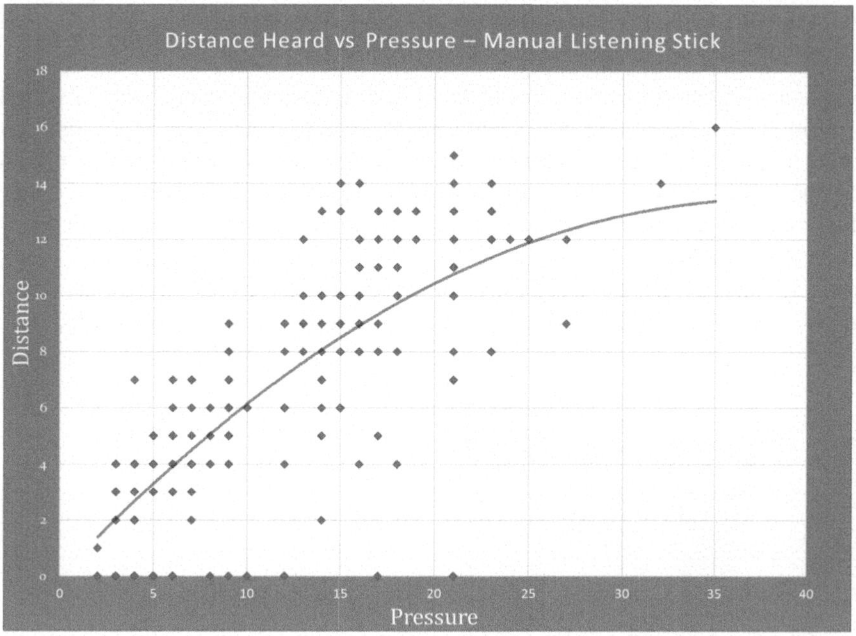

Figure 7.3 Plots of leaks found using a manual listening stick (Source: Stuart Hamilton).

noise is not being amplified so the operator does not hear it. Another factor could be that, with a listening stick, background noise cannot be eliminated as it can be when using the headphones used by electronic device operators.

What can also be seen is that many leaks could not be heard at all by the individual and yet were heard when transferring to the electronic device. This could mean that, when conducting a survey using a manual stick, many leaks could be missed.

If the person listening has poor quality hearing, then a leak may not be heard until the operator is much closer to it where the leak noise is louder, whereas someone with better hearing could potentially hear further away. The possibility of hearing quieter leak noises is greatly increased if using an amplified electronic listening device.

When using either a manual or electronic listening device, if the frequency range of a leak noise is below or above the leak detection engineer's hearing range, then that leak will not be heard, regardless of loudness.

Based on all this, it is recommended that an electronic listening device should always be used regardless of the level of pressure, since using a manual stick reduces the ability to hear leak noise. Of course, electronic devices such as leak noise correlators or noise loggers can be used and may hear leak noises at low frequencies or low sounds but experience shows that the distance these devices can locate a leak using magnetic connections on accelerometers, are still comparatively low compared to when they are used on metallic pipes.

Many plastic and polyethylene pipes have different noise transfer abilities and with experience it is found that older pipes, that may have hardened due to age, have a better noise transfer ability than newer softer pipes.

7.8 WORKING IN LOW PRESSURES SYSTEMS WITH PLASTIC AND POLYETHYLENE PIPELINES

Figure 7.4(a–c) shows engineers checking for leaks in different situations in a system with 4 m pressure. The engineers were always aware that they would not be able to hear a leak more than 4 m in distance from the listening point. The leakage engineers used an electronic listening device to listen on house connections, be they either a tap or meter, to conduct the leakage detection exercise. This was because the distance between the fittings to be listened to were less than 4 m apart, so that the whole pipe could be checked acoustically.

In both Figures 7.4 and 7.5, the distance between the service connections is less than 4 m, meaning that, when listening on each service connection, leaks on the main pipe would also be heard.

Figure 7.4 (a–c). Checking for leaks. Note: lines indicate the location of the pipework and the arrows indicate where the service pipe enters the property. (Source: Stuart Hamilton).

Figure 7.5 (a,b) Checking for leaks. Note: lines indicate the location of the pipework and the arrows indicate where the service pipe enters the property. (Source: Stuart Hamilton).

In Figure 7.5(a, b) the pressure is at 4 m but distance of the length of pipe between the service connections is more than 4 m, therefore sounding with an electronic ground microphone every 1 m above the main pipe had to be done to detect any leaks that could be present but cannot be heard when listening on the service pipe or house connections.

During night time (Figure 7.6), pressure had increased to 7 m but, since service connections were inaccessible (because the water meters were inside the properties), sounding with an electronic ground

Figure 7.6 (a,b) Checking for leaks at night-time. (Source: Stuart Hamilton).

microphone every 1 m above the main pipe therefore had to be done to detect any leaks on the main pipe to the service connections.

7.9 CONCLUSION

Based on the finding that leak noise will travel 1 m in distance for every 1 m in pressure up to 15 m in plastic and polyethylene pipes, leakage detection exercises can be carried out using an electronic ground microphone even in (very) low pressure systems.

The survey approach is to listen on house connections and on mains fittings, provided that the total distance between the fittings does not exceed the distance that the noise may travel, so that all the pipework is checked acoustically. If the distance between the fittings is greater than the pressure in the main – meaning that some of the pipework would not be checked acoustically if listening only on the service pipe or house connections – then the pipework needs to be checked by listening on the ground surface above the route of the pipe in 1 m intervals.

The ability for leak noise to travel to the surface is not so much reliant on pipe material. However, it is reliant on other characteristics such as:

- pressure in the pipe;
- type of leak;
- size of leak;
- type of ground backfill;
- road or footpath surface above (used to listen on);
- moisture in the ground between the leaking pipe and the ground surface;
- water table variants, if close to the sea (high tides).

For these reasons, failure to be directly over the route of the pipe may prevent the leak noise from being heard.

When conducting a survey with an electronic device, the ability to hear a leak present in relationship to pressure and distance is nearly linearly proportional until around the 15 m point. In our studies, the maximum distance that a leak could be heard was 28 m from the source of the leak to the service or house connection, with a pressure of 45 m.

When conducting a survey using a manual listening stick, the numbers are not at all proportional and can be considered quite scattered, with the results being considerably poorer than when using an electronic device. In our studies, the maximum distance that a leak could be heard using a manual listening stick was 16 m from the source of the leak to the service or house connection, with a pressure of 35 m.

Based on the above, the benefit of using an electronic listening device is clearly obvious. When using either a manual or electronic listening device, if the leak noise frequency range is below or above the threshold that an individual can hear, then this leak will not be heard, regardless of how loud it is.

There is evidence that, when doing leakage surveys on metallic distribution systems by listening on mains fittings only where rehabilitation work has been completed using plastic tails on valves, or where fire hydrants or repairs using plastic and polyethylene piece throughs are present, the ability to hear leak noise is greatly affected and the approach should be changed to also listening on all house connections.

Electronic devices such as leak noise correlators or noise loggers can be used and may hear leak noises at low frequencies or low volumes but experience shows that the distance these devices can locate a leak using magnetic connections are still comparatively low compared to when being used on metallic pipes.

Many plastic and polyethylene pipes have different noise transfer abilities and experience has shown that older pipes which may have hardened due to age have a better noise transfer ability than newer softer pipes.

On all occasions it is recommended that an electronic listening device is used to achieve better results, compared to the use of a manual stick. Leakage engineers should be checked for excellent hearing capability so that it is established from the outset that they can hear any leaks present. All work in busy areas should be completed when the background noise is at its lowest.

Network conditions will dictate the procedure and approach to be adopted for a leakage survey. The ALC survey approach should be adapted depending on many network characteristics and should not only restricted to the following:

- day or night working;
- pressure in the main;
- distance between fittings;
- rural or urban regions;
- intermittent supply;
- water table variants if close to the sea (high tides);
- mains materials;
- service connection materials;
- road or footpath surface to listen on;
- ability to gain access to house connections.

7.10 ACKNOWLEDGEMENTS

The surveys and findings presented in this chapter were carried out on projects around the world with contributions and assistance from the following:

- Roland Liemberger (Austria); roland@liemberger.cc
- Tomislav Vitovski (Croatia); tomislav.vitovski@aqualeak.hr
- Miya Water (Spain); www.miya-water.com

Chapter 8

Losses and how to find them – rules that apply to all situations

8.1 METHODOLOGY FOR LEAKAGE DETECTION ACTIVITY

All staff must develop the experience and specialist skills to properly assess the differing methods of leakage detection, and thus choose the appropriate method depending upon area site characteristics such as access, operation and level of leakage.

Before heading out into the field, staff should collect and analyse all available information from within each network area. This collected information will inform the choice of the appropriate leak detection methodology, taking into account certain site characteristics such as:

- geographic area;
- mains length;
- customer type;
- number of properties;
- pipe materials;
- asset condition;
- water supply period;
- water pressure in the region to be surveyed.

Figure 8.1 shows a graphical representation of this analysis process. It is a leakage detection determination methodology, formatted as an easy-to follow flow diagram.

Starting at the top of the figure, each part of the flow diagram is described step by step as follows:

- **Characteristics: Geographic area.** In advance, understand the geographic area where the crew will be working and in what surroundings, know what time of day the work will be completed, and know what is required to complete the work safely. This will affect what leak detection approach should be adopted and what technology should be used. Understanding this at the outset will avoid arriving on a

Author(s): Stuart Hamilton, Bambos Charalambous and Gary Wyeth
doi: 10.2166/9781780409207_0069

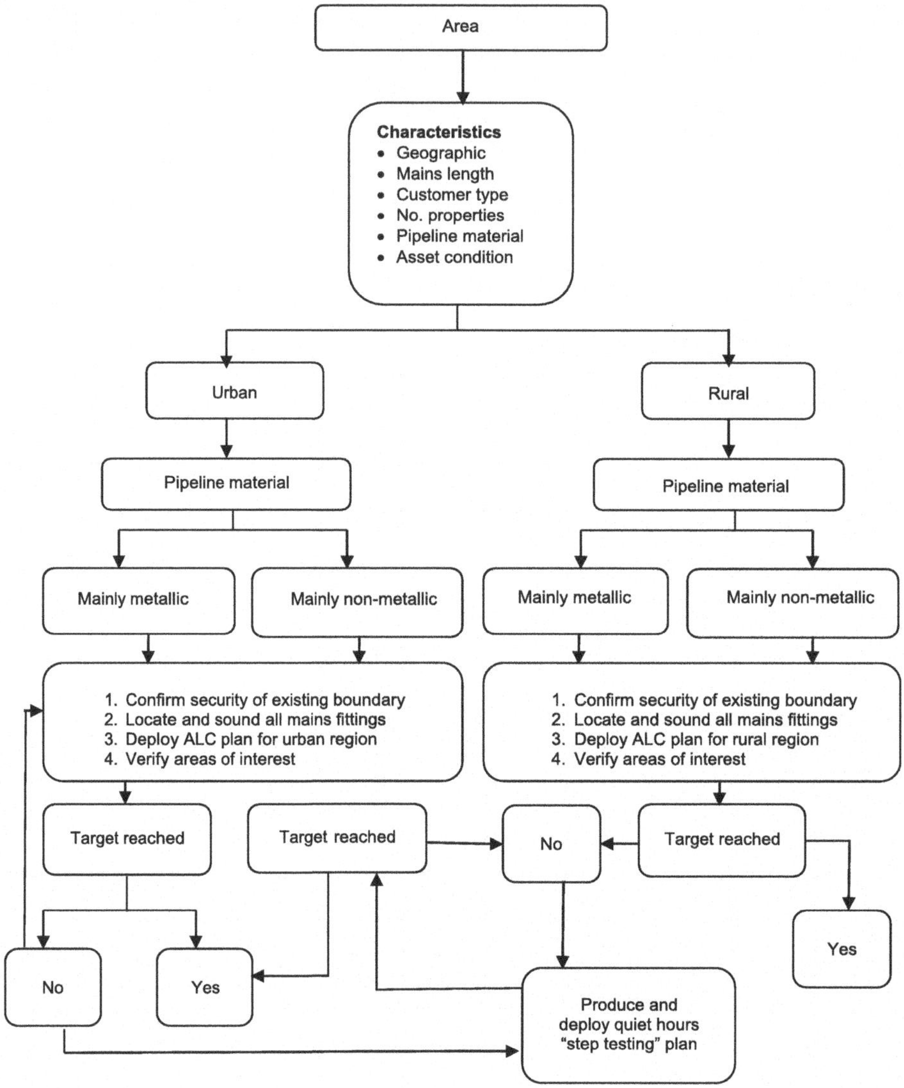

Figure 8.1 Standard leakage detection methodology.

site with no information, and thus the crew will be prepared for the local situation that they will encounter.

- **Characteristics: Mains length.** Determine in advance the length of the mains in the area, as that will determine how long it will take to complete the leak detection survey, and thus the time scale it will take to complete work in the area.
- **Characteristics: Customer type.** Gather information in advance on the characteristics of the customers in the area where the leak detection survey will be conducted. Is the customer type commercial or domestic? Is the housing large or small, apartments or houses? Is the area wealthy

or poor? This information helps to properly plan the leak detection approach needed prior to going on site. If the area is commercial, conduct meter readings on large customers prior to completing the leakage survey. Note any large customers in the middle of a domestic region (such as hospitals or hotels) that may consume large quantities of water 24-hours a day. Noting these large customers as causes of high usage in advance will prevent them from being mistakenly considered as leakage later.

- **Characteristics: Number of properties.** Make an inventory of the number and type of properties in the area of the leak survey in order to determine the number of fittings available to listen on.
- **Characteristics: Pipe material.** Knowing the pipe material content in advance – along with the pressure – is critical, as there is a different approach to leakage in metallic and non-metallic regions.
- **Characteristics: Asset condition.** Obtain information on the condition of the network asset in the area to be surveyed, such as age, number of prior repairs, and any other information that might be available.
- **Water supply period.** Understand the period that the water supply system is pressurised (in the case of intermittent systems) and to what pressure the system is operating on. Ensure that all surveys are conducted during the water supply period and after it is considered that all tanks are filled and the mains pressure is at its highest.

A successful leak detection plan requires knowledge of the above characteristics of an area to enable proper planning in the office prior to going on site. This information will determine leakage location activity, and the length of time needed to conduct the survey based on the number of connections per day and/or length of mains in the area. Knowing the distance between fittings and the pipe material, along with pressure, will also help to inform what approach to undertake and what technology will work best. Knowing the housing type will allow you to consider consumption per property, which will assist in working out whether the problem is high usage or leakage. Also, knowing the extent of commercial activity in the area will dictate as to when to conduct the leak survey.

Continuing through the diagram:

- **Urban or rural?** If the area is rural with a long length of mains but with few properties, conduct a Step Test rather than walking the mains. Similarly, if the area is urban with heavily populated properties, then night time leakage detection may be the better solution. However, before deciding to work at night, conduct a daytime visual inspection to understand where fittings are and where the water mains run.
- **Material content: Mainly metallic.** If the system is mainly metallic, consider the distance between fittings. Acoustic noise will travel further in a metallic system, but the noise will not travel long distances. Acoustic technology such as correlators will work easier in these circumstances.
- **Material content: Mainly non-metallic.** In non-metallic systems, listening to fittings will not work. The noise of any leak will not be easily heard since the energy created by the leak is absorbed by the non-metallic material. In these circumstances, acoustic devices such as correlators are of poor use, and other approaches need to be considered. ALC for leaks (listening above the pipe at ground level) is one such suitable approach.
- **Confirm security of existing boundary.** Ensure that all boundary valves are in a closed position and that no water is leaving the area to be surveyed. This is important, as the indication of high leakage may be from a valve being open and not from an actual leak.
- **Locate and sound all mains fittings.** Locating all fittings during daylight hours is imperative, as many hours can be lost at night looking for a valve to conduct a listening survey or step test.
- **Deploy ALC plan for urban region.** Understanding the type of region will determine which methodology will provide the best chance of quickly locating the problem. The sooner a problem

is located, the sooner it can be rectified. Urban region characteristics such as heavy traffic, high population, and proximity to schools and commercial businesses will mean that the time of day to conduct leak surveys must be planned in advance.

- **Deploy ALC plan for rural region.** In rural regions, fittings are few and the distance needed to be surveyed is high, so either a drive through the area as a first instance or conducting a step test to narrow down the problem should be considered.
- **Verify areas of interest.** Not all findings will be leaks, so initially these should be classified as 'areas of interest' (AOI) where the technician must verify by all possible means that the AOI is actually a leak. Because of the high costs and time frames associated with excavating, 'dry holes' or aborted excavations (when leaks are not found) must be avoided.
- **Target.** Determine the volume of leakage to locate and inventory the volume of the individual leaks that have already been located. This is necessary to determine if the target of the leakage volume has been reached. Consider withdrawing from the region until the leaks have been repaired and a new measurement taken. If the exit target has been reached, then move on to another area.
- **Produce and deploy a quiet hours 'step testing' plan.** Produce a step test plan to carry out the valve closing exercise at a time when most convenient, when consumption is at the lowest in the region. Not all step tests have to be completed at night. Many, such as those in rural areas, can be completed during the day.

Part Two

CASE STUDIES 1–10

The ten Case Studies presented here highlight the application of the ladder concept and how companies around the world have managed to move up the ladder from intermittent to 24×7 supply regimes and/or delivering different approaches to Active Leakage Control depending on their system's characteristics.

Author(s): Stuart Hamilton, Bambos Charalambous and Gary Wyeth
doi: 10.2166/9781780409207_0073

Case study 1

Five cities around the world

Stuart Hamilton (United Kingdom)

CASE I – CITY IN THE MIDDLE EAST

This area was on intermittent supply and there was a plan to move the area to 24-hour supply to conduct leakage detection. Due to being intermittent 4 hrs per day every 3 days, the householders wanted to ensure that they had their water tanks full and since they were located on top of the highest part of the properties, these tanks contained no shut off valve such as a ball valve, for example, so customers could see that a tank was full when it over-flowed.

A programme was started to install control valves to these tanks so that the 24-hour supply scheme could be introduced and a team of plumbers was sent into the region to install ball valves and taps to open ends. The programme installing control valves was in full swing but at the end of the day the street was littered with taps and ball valves which the customers had removed.

The customers had spent many years needing to be sure that the tanks were full and overflowing to be sure they had water, so to remove this made them nervous, and they wanted to go back to the old ways. ALC was therefore postponed until a complete news and education programme could be introduced showing the benefits of controlling water losses. Fitting of control devices was restarted some months later, finally allowing an ALC programme to take place.

CASE II – CITY IN THE MIDDLE EAST

In this case, the same situation was initially experienced as in Case I but, this time, to move forward with an ALC programme all services connections were located and turned off on a day when customers were not expecting to get any water. The mains distribution system was pressurised, allowing mains side leakage activities to take place and leaks to be located, ready to be repaired at a later date.

CASE III – CITY IN ASIA

The city was supplied by a very low pressure system with pressure achieving less than 5 m pressure, 0.5 bar or 7 psi, and it was practically impossible to locate leaks with acoustic methods. What was tried with success in this area was to locate areas of leakage by the use of a step test locating the street that had a problem and then isolating the street from the rest of the system. With the use of a tanker lorry, water was pumped into the street via a fire hydrant to pressurise the street pipes to enable the leak to be located via acoustic methods. This was an extremely difficult exercise but did allow the leak to be located with success.

CASE IV – CITY IN THE MIDDLE EAST

This city region was extremely hot and water was scarce, with intermittent supply being the standard situation. The city network was in relatively good condition in terms of mains and service fittings, and each property was fitted with a working control valve.

Conducting acoustic leakage investigation was deemed to be impossible so the alternative of gas injection into the dry water mains was used. The process involved turning off all property connections and areas of the system were enclosed to control the area being worked on. Gas was injected along with air from a compressor and the leak finding process was conducted. The process was successful in finding many leaks; however, it is not suggested that this would find every leak in a water system. This was a successful method for finding leaks in a dry system.

CASE V – CITY IN AFRICA

Pressure in this city was low and the system was relatively old with very few house connections visible. It was decided that a visual leakage survey should be conducted in the first instance. During this visual survey, turning the water on produced numerous leaks and open ends, all of which were discharging water as leakage. On average, a leak was located every 100 m and this process should be adopted in many cities and regions in the first instance, prior to the introduction of technological solutions, as it is cheap and can be conducted with limited training, and will discover many leaks for repair.

CONCLUSION

To conduct leak investigation activities in intermittent supplies using acoustic methods, the system has to have some pressure in it to transmit the energy from the leak along the pipeline to be heard. Water has to be in the system long enough to fill all tanks and each tank has to be controlled. Additionally, all properties should stop drawing water or risk interfering with the process and giving false indications that properties have potential leaks, resulting in them being investigated unnecessarily and hence causing ALC to be a long process.

The method of gas injection outlined in Case IV above is an alternative but is not necessarily cheap or readily available, and a significant amount of gas is required to cover a system and experienced operators are required.

The alternative of visual leak investigation requires no experience and no technology, and is probably always the first approach to be adopted and should only be carried out when the water is turned on.

Case study 2

A holistic approach to system improvement

Miya Water/Bahamas Water and Sewerage Corporation (The Bahamas)

This case study outlines the preparation and implementation process of the Bahamas Water and Sewerage Corporation (WSC) and Miya Water (Spain)'s performance-based contract (PBC) for non-revenue water (NRW) management in New Providence, Bahamas. This innovative contract was implemented as part of the International Development Bank (IDB) Loan Project: WSC Support Programme – New Providence Water Supply and Sanitation Systems Upgrade.

BACKGROUND

In 2012, WSC contracted Miya Water to conduct a 10-year NRW reduction project. As a part of this contract, Miya Bahamas was required to reduce NRW in the New Providence distribution system from 6.87 Million Imperial Gallons (MIG) to an annual average of 2.5 MIGD (Million Imperial Gallons per Day) by year 5, and to 2.0 MIGD by year 7 (1 MIG = c.4546 m^3).

The island of New Providence, on which this programme is centred, accounts for approximately 70% of the population of The Bahamas (351 000 inhabitants). The availability of renewable freshwater per capita in the country, and on this island, is one of the lowest in Latin America and the Caribbean.

For this reason, the island population relies on groundwater and desalination as its only two drinking water sources. Groundwater, however, is very limited and over-abstraction is a serious concern. The well fields in New Providence have a maximum estimated capacity of 1.5 MIGD, while customer demand fluctuates at around 10 MIGD.

For this reason, over 90% of the drinking water supplied to the island comes from reverse osmosis plants, which yield a comparatively expensive product. In this context, the water losses in the distribution network, estimated to be close to 50% of the water produced, were considered economically and financially unacceptable. These losses originated from leaks in the network, unauthorised consumption and metering inaccuracies.

PERFORMANCE

A baseline survey report showed that the baseline of NRW in New Providence in 2011 was 6.87 MIGD at an annual average system pressure of 24.6 psi and with a pressure-leakage exponent (N1) of 0.9. The leakage exponent (N1) shows the interdependency of leakage on pressure. N1 values are typically found to range from 0.5 to 2.5, depending on pipe material and level of leakage. Miya designed an NRW reduction strategy to ensure that all the contractual targets were achieved, starting from this baseline level.

The NRW reduction strategy was designed to meet and maintain the target level of NRW of 2.5 MIGD at an annual average pressure of 25 psi by 2018, and to further reduce and maintain the target level of NRW at 2.0 MIGD by 2019, maintaining this level until 2022.

The NRW strategy included details of the condition, performance, operation and maintenance of the WSC distribution system assets in New Providence, and their impact on NRW management, in order to provide a clear understanding of the reasons for various components of the proposed strategy. The report also detailed the leakage modelling undertaken using IWA methodologies, including a component analysis of breaks and background losses, an economic analysis of the various NRW reduction strategies considered and details of each component of the proposed NRW reduction strategy.

The key IWA performance indicators for the 2011 annual water balance were:

- NRW as a percentage of System Input by Volume (SIV): 57.8%;
- NRW as imperial gallons/connection/day: 147;
- apparent losses as imperial gallons/connection/day: 9;
- real losses as imperial gallons/connection/day: 136;
- Infrastructure Leakage Index (ILI): 27.5.

It can be seen from this data that 92% of the total NRW volume was attributable to real losses. The NRW reduction strategy was therefore primarily focused on reducing real losses.

The average system pressure of 24.6 psi in 2011 was a significant increase over the estimated 19 psi estimated by WSC during the tender period. This increase in system pressure is certainly one of the reasons that real loss volumes increased, but it does also mean that there was greater scope for pressure management to reduce real losses and break rates at the start of project execution.

UNIQUE PROGRAMME FEATURES THAT DROVE THE SUCCESS

A number of unique features ultimately drove the programme's success:

- it was an output-driven contract: the deliverable was the NRW level to be reached and later maintained, not a bill of quantities or specific civil works. This allowed flexibility and focus on achieving the goal while minimising unnecessary works;
- a quality-based selection process ensured quality of design and implantation;
- a high performance-based fees component of 30% directly related to the actual achievement of the targets created an alignment with the utility;
- the high performance-based component also served as a performance guarantee (thus only a small guarantee was required for the implementation period);
- the turn-key holistic approach and the high performance-based fees component ensured full accountability of Miya and their striving for best results;
- as a contribution to the community, the project included a water efficiency education programme in local schools, including implementation of water conservation measures.

METHODOLOGY

The NRW reduction strategy concentrated on reducing real losses. However, the volume of apparent losses was monitored over the course of the contract and if apparent losses increased above the economic value, work would have been undertaken as necessary to reduce apparent losses to an economic level.

Current industry best practice for reducing real losses revolves around implementing the four IWA basic methods of managing real losses. The NRW strategy utilised all four of these methods:

- **Component analysis of breaks and background losses: A Breaks and Background Estimates (BABE).** BABE component analysis disaggregates the overall volume of real losses determined from the IWA annual water balance into components, so that it is possible to determine how much of the overall real loss volume is attributable to the different components of the distribution system.
- **Analysis of pressure zone nightlines and total daily flow.** The IWA annual water balance determines the overall volume of NRW and how this volume is broken down into real and apparent loss volumes, but it does not quantify the losses by area in order to inform the NRW reduction strategy.
- **Geographic analysis of historic break data.** The project team undertook a geographic analysis of the historic work management data that WSC holds on mains and service line breaks in order to identify the sections of mains with the highest break rates per kilometer and roads with the highest service line break rates per kilometer. During the course of the implementation phase, over 17 000 service connections were replaced according this study.
- **Economic analysis**. An economic analysis of all suitable NRW reduction options was undertaken in order to develop the optimal least cost NRW reduction strategy for New Providence. The analysis showed that the economic intervention frequency (the interval between full comprehensive surveys, if undertaking a system-wide survey) was 5.5 months. Therefore, the pace of leak surveys in miles to be performed during each year was determined.

However, the NRW reduction strategy has been regularly refined during project execution as better data is collected and experience and data is gained from implementation of the components of the strategy (Figure CS2.1). The annual targets for NRW reduction and average annual pressure are detailed in Table CS 2.1.

Figure CS2.1 Data loggers gathering system data on flow and pressure (Source: Miya Water).

Table CS2.1 The annual targets for NRW reduction and average annual pressure.

Annual Average Target (Calendar Years)	Year 1 2013	Year 2 2014	Year 3 2015	Year 4 2016	Year 5 2017	Year 6 2D18	Year 7 2019	Year 8 2020	Year 9 2021
NRW (Migd)	6.0	5.0	4.0	3.0	2.5	2.5	2.0	2.0	2.0
NRW (g/conn/d)	129	103	79	57	46	45	35	33	32
Average Annual Pressure (psi)	20	21	22	23.5	25	25	25	25	25
Agreed Pressure Adjusted NRW Reduction Target (Migd)	0.45	1.38	2.34	3.36	3.97	3.99	4.51	4.53	4.56

Source: Miya Water.

The Project is composed of two 5-year phases. In the first phase, the implementation phase of the project, several activities were undertaken by Miya including the following:

- number of service connections replaced = 17 077;
- number of service connections disconnected = 5724;
- number leaks repaired = 5640;
- number of PRVs installed = 32;
- number of boundary controls installed = 245;
- number of DMAs established = 89;
- number of new pressure tappings installed = 180.

The second phase of the project, 2017–2022, is aimed at the maintenance of the NRW assets already installed, and on pressure management, maintenance of DMAs including boundary control, and large customer meter maintenance and active leak detection, among others. Highlights include:

- miles of main covered by active leak detection (until March 2020) = 12 937;
- number of large customer meters replaced (until March 2020) = 77;

Figure CS2.2 Leakage detection in progress (Source: Miya Water).

- number of bulk meters refurbished/replaced (until March 2020) = 106;
- number of system valves cleaned/exercised (until March 2020) = 2723.

CONCLUSIONS

The NRW reduction levels achieved between 2012 and 2017 were significant and constitute the single most important accomplishment of the programme. They reflect the overall success of the NRW reduction strategy. In 2019, the NRW level was reduced to 2.00 MIGD, once average water pressure and the impacts of delays on leak repairs were taking into consideration.

The key elements of the strategy were proactive leak detection, rapid repairs, use of adequate materials, pressure management, selective replacement of network elements, disconnection of inactive service lines, large customer metering, and asset maintenance. Other essential components were hydraulic modelling, system optimisation, GIS updating, SCADA, and the use of data management hardware and software (NETBASE).

The economic impacts of the NRW reduction programme are significant. Annual cost savings were achieved because of a reduction of the system input volume purchased and the increases in revenues compared with the baseline. At the end of 2018, the cost savings due to reductions in system input volumes had already amounted to US$31.5 million, based on marginal production costs of US $8.18/1000 gallons.

The increase in revenue, compared to the baseline, amounted to US$37.2 million Thus, MIYA estimates that the total financial benefits amounted to US$68.7 million by 2018.

Case study 3

NB-IoT expedites leak detection

Envirotec Magazine (7 September 2020)

Yorkshire Water and partners are nearing completion of the deployment of almost 4000 acoustic, flow, pressure and water quality monitors in what is said to be the UK's largest smart water network pilot. Final installations of the latest NB-IoT (Narrow Band Internet of Things) pilot technology were recently underway, the fruit of a collaboration between British Telecom, Yorkshire Water and Stantec after British Telecom switched on its first upgraded masts in the UK, as part of the project. If the pilot is successful, said the utility, 'the smart water network will revolutionise the way leaks and interruptions to supply are managed in the future.'

British Telecom's NB-IoT solution has the potential to deliver significant improvements in data quality and battery life, allowing the utility to identify and prevent leaks and network incidents more accurately than before.

Figure CS3.1 Final installations of the latest NB-IoT (Narrow Band Internet of Things) pilot technology were recently underway (Source: Envirotec magazine).

The Smart Water Network Pilot will integrate the data from multiple new and existing sources and present it in a single visualisation platform which will include a digital twin of the water network. The platform will use AI to cluster data sets, and remove false positives, to accurately inform asset and operational decision making.

Yorkshire Water's chief strategy and regulation officer, Nevil Muncaster, said: 'We recognise that on top of our commitments to our customers and the environment, we need to play our part in regional development where we can – perhaps even more so than ever in these challenging times. It's great that through this innovative pilot we've been able to accelerate the introduction of NB-IoT to the area in Sheffield. NB-IoT has the potential to provide greater access for local businesses to take advantage of the advancement in IoT technology, which can only be a good thing for economic growth within the region.'

As part of the Smart Water Network Pilot, Gutermann acoustic loggers, Technolog pressure loggers and Honeywell flow meters will use the NB-IoT network to transmit their data.

Sarah Walker, director for British Telecom's Enterprise business in the North of England said this partnership marked an exciting milestone for the company. 'The pilot deployment of NB-IoT signals a move to a more data driven world enabling millions of connected devices to send and receive data, transforming the way we live and work in the future.'

Case study 4
Anglian water

Ovarro Ltd/Anglian Water (UK)

THE CHALLENGE

Anglian Water, a UK-based water company operating in the east of England, has reduced leakage by 20% since 2010 and has committed to a further 23% reduction by 2025. Overwhelmingly, customers have told Anglian Water that leakage is a high priority and that they want to see it reduced to well below the Economic Level of Leakage (ELL).

The aim was to develop, own and utilise a system that enables cost effective, smart, real time leakage monitoring at an individual asset level for Anglian Water's entire network. The team were faced with a multitude of challenging conditions: the sites consisted of a large range of materials including but not limited to cast iron, ductile iron, asbestos cement, MDPE and PVC.

THE SOLUTION

Trials were undertaken in Lincolnshire, England, in order to compare different technologies over a total mains length of 40 km, comprising of mixed material pipes. 800 traditional older style noise loggers were deployed every 50 metres. By comparison, 58 Enigma3hyQ units were deployed every 750 metres to cover the same area – this represented a 65% cost saving when compared with traditional noise logging.

The results given by the Enigma3hyQs were demonstrably much better, with all units communicating, fewer 'false positives', zero theft and lower leakage levels enabling a faster response time with more positive leak positions identified allowing the continued drive to reduce leakage levels.

ADVANCED PERFORMANCE

The Enigma3hyQ units used offer a number of benefits:

- optimised location over longer distances, plastic and large diameter trunk mains;
- accurate time synchronisation (patent #2555053);
- no above ground radio required;

- remote leak listening for confirmation of leak noise;
- multi-point correlation via 3G/GPRS communication.

KEY DELIVERABLES

After the trial, Anglian Water installed 2350 units of Ovarro fixed network loggers. And after more success in reducing leakage, they installed a further 3500 units of Enigma3hyQ. Despite the challenging conditions, the Enigma3hyQs continue to pinpoint leak locations accurately over distances of up to 4.6 km.

Overall, by deploying Enigma3hyQ loggers, Anglian Water found 1325 leaks initially and now regularly see leakage reductions of 1.4 million litres of water per day in targeted areas, helping the water company to meet regulatory targets and support them in reducing leakage.

Case study 5

Ranhill SAJ Sdn Bhd

Ovarro Ltd/Ranhill SAJ Sdn Bhd (Malaysia)

THE CHALLENGE

Ranhill SAJ Sdn Bhd, a subsidiary of Ranhill Holdings Berhad, is an integrated water supply company in Malaysia, involved in the process of water treatment and distribution of treated water to consumers, right up to billing and collection. It serves a population of around 3.1 million and manages 22 175 km of pipes over an area of approximately 19 000 km^2.

The Non-Revenue Water (NRW) of Johor, as at December 2018, was 24%. To achieve a target NRW of 5% by 2025, several activities have been undertaken. One of the activities implemented to achieve the target was the creation of SMART DMAs. In the initial stage, five DMAs were involved in implementing the SMART DMA programme.

ENIGMA3 m noise loggers were deployed at a DMA for at least three months: in the first month to find leaks from the correlations obtained, in the second month whilst leaks were repaired, and for monitoring purposes during the third month.

THE SOLUTION

The total pipeline length to be covered for the Bandar Putra B DMA was 38.51 km with 5052 connections, whilst for the Bandar Putra B DMA it was 38.51 km with 5052 connections.

The net night flow (determined by subtracting legitimate night flow from the minimum night flow) before installation of the ENIGMA3 m loggers was at 30.99 l/s and the total daily flowrate was 6200 m^3/day.

During the implementation of SMART DMA, a total of 295 ENIGMA3 m units were installed. Pipeline material and the distance between ENIGMA3 m loggers varied depending on pipe material (from metal pipe to poly pipe).

A total of 115 leaks were found and repaired in the second month of ENIGMA3 m installation.

KEY DELIVERABLES

Further to the results obtained from ENIGMA3 m installation at Bandar Putra, the test was also successful in that small leakages at hydrant valves and communication pipes were also able to be located.

Overall, the net night flow after 3 months of ENIGMA3 m installation was reduced to 20.08 l/s from 30.99 l/s. This reduction gave a saving of 705 m^3/day with a total current water cost saving of US $4000/month.

Case study 6

Acoustic leak detection solution delivering from day one

Kamstrup / Söderhamn Nära AB (Sweden)

DELIVERING FROM DAY ONE

Söderhamn Nära AB (a Swedish water utility), went from a costly and time-consuming hit-and-miss approach to leak detection to a targeted effort with their new acoustic leak detection solution. Whilst being part of a 4-year plan moving towards hourly values and analytics, the new solution literally delivered from day one and is still providing valuable insights.

Every year, 400 000 m^3 of water disappears on its way to Söderhamn Nära's customers – the equivalent of a 20% water loss. In an effort to bring that number down, the utility started rolling out a new smart metering solution that makes it possible to detect leaks in pipes via acoustic leak detection.

According to Robin Lindberg, Project Manager for Söderhamn Nära AB, leak detection has always been a laborious and lengthy task. If they suspect a leak, they go from valve to valve to listen for it, shutting off areas one by one to verify their assumption. Areas with plastic pipes are particularly problematic. 'In plastic pipes, you can't hear the leak with our traditional equipment so, unless it comes up to the surface, it's almost impossible to find.' Water loss is especially challenging in the city. Leaks here mainly occur in the bigger pipes and can be as much as 30 m^3 per hour. 'Finding a leak can take anywhere from days to years,' says Henrik Inancsi, Meter Manager for Söderhamn Nära AB.

Today, however, they use their new flowIQ® 2200 smart meters combined with the analytics tool, Leak Detector, to identify the most relevant addresses and then go there to verify potential leaks. As a result, their efforts are now a lot more targeted, which ensures an efficient use of resources.

DAILY OPERATIONS TOOL

Robin Lindberg and Henrik Inancsi both experienced a steep learning curve working with Leak Detector, which is now a part of their daily operations. They currently have 15–20 addresses on their watch list where they follow developments in noise levels closely. 'We have learned to tell the different noises curves apart. In the beginning we went out to check all meters that detected an unexpected noise. Now

Figure CS6.1 Robin Lindberg and Henrik Inancsi both use their new flowIQ® 2200 smart meters to identify potential leaks.

we can see from the noise pattern if it is an actual leak or something else.' In several cases, the culprit turned out to be a heating pump or even a dehumidifier in a laundry room.

At just 5–10 litres an hour, one leak in a toilet was too small to be picked up by the normal leak alarm in the meter. But because the acoustic leak detection in the new meters is very sensitive, the leak was found and the utility could inform the customer. At the other end of the scale, the water consumption at one address suddenly went from 0 to around 5000 litres per hour one Saturday evening. The customer was informed Monday morning and subsequently discovered an open valve that had resulted in 300 m^3 lost in one and a half days.

LOCAL HEROES

The new meters have found about 100 leaks inside homes and properties in Söderhamn according to Henrik Inancsi. 'So far we've made 50–60 calls about incidents like leaks, running toilets and garden hoses that have been left on. They are all based on info codes from READy [the meter communication package] or noise patterns in Leak Detector.' He says that some customers are suspicious at first because they have not noticed anything out of the ordinary but once they have checked they often call back and are very grateful. 'Not only do we prevent potential collateral damage from a long-lasting leak. They also avoid paying for water they don't use. This is the basis for great customer relations, which is so important to us.'

The new solution has also attracted a great deal of positive attention from colleagues in the Swedish utility business and media. 'A lot more people are talking about it and asking questions now than 6 months ago – and we're happy to share our experience, because we know this works,' says Robin Lindberg.

Figure CS6.2 The flow meter data is uploaded to a laptop using the READy programme.

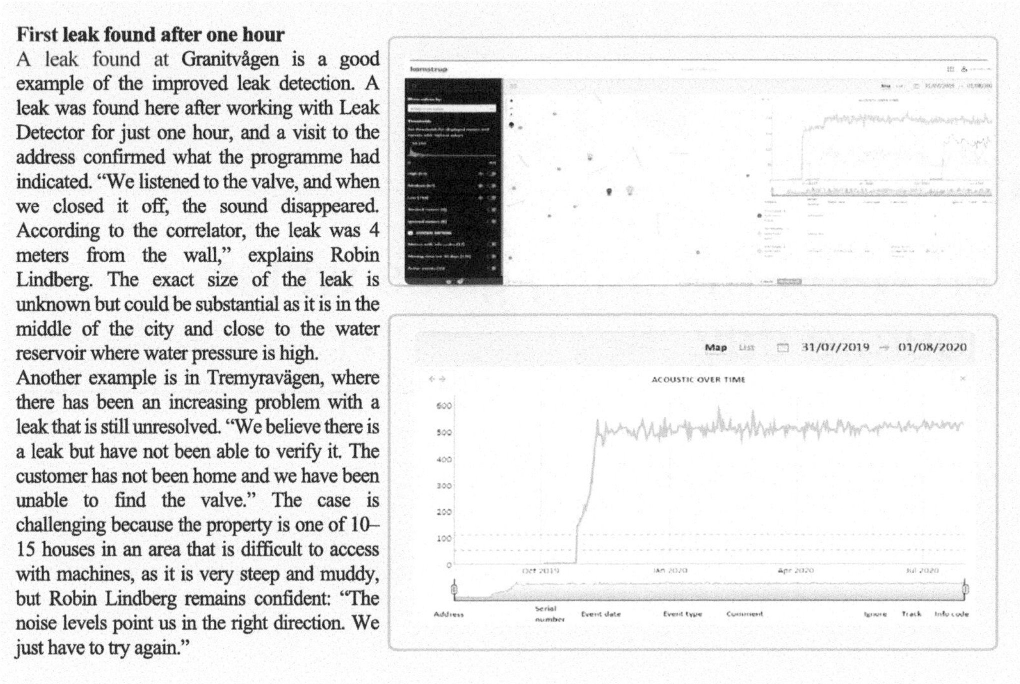

Figure CS6.3 Leak Detector in operation.

ON THE RIGHT TRACK

For Söderhamn Nära AB, focus right now is on replacing the last of the meters, which is the next step on their way to the goal they expect to reach within 3–4 years. 'Our goal is to have hourly measurements from all meters and to add district meters as well, so that we can implement more analytics in a few years,' says Robin Lindberg.

Another example is in Tremyravägen, where there has been an increasing problem with a leak that is still unresolved. 'We believe there is a leak but have not been able to verify it. The customer has not been home and we have been unable to find the valve.' The case is challenging because the property is one of 10–15 houses in an area that is difficult to access with machines, as it is very steep and muddy, but Robin Lindberg remains confident: 'The noise levels point us in the right direction. We just have to try again.'

When all meters deliver hourly data, that enormous amount of data will provide a much better basis not just for leak detection and customer service but also for maintenance and asset management investments. 'The more meters we install, the more we'll know about trends and seasonal variations and the more effective the system will be. That is when we will see the full value of our data. We're a small utility so we have to take it step by step, but we're definitely on the right track.'

Case study 7

Smart water metering gives water 24-hours a day

Kamstrup/Maharashtra Jeevan Pradhikaran (India)

Water is in short supply in many cities all over India. After installing smart meters in one city, customers now have access to water 24 hours a day.

EFFECTIVE WATER MANAGEMENT

In 2015, 450 water meters were successfully installed in the city of Kapil Gram Panchayat, 300 kilometers southeast of Mumbai. The city has a population of 5000 people and was the first city in India to get ultrasonic waters meters installed.

Before getting a Kamstrup meter, consumers were charged at bulk rate. There were no limitations on how much water could be consumed and how much wasted. By adding a metering system, consumers are now aware that it is better to consume less and pay less. In other words, they save costs that directly impact their pocket.

Kapil Gram Panchayat decided to install AMR water meters for effective water management. With the project now running, the expectations remain high, according to Mr S. K. Bhopale, Sectional Engineer, Maharashtra Jeevan Pradhikaran. 'To recover project implementation costs, we needed to issue water bills to all on a monthly basis. And the bill is most important, which should be based on actual usage of each consumer. That's why we installed Kamstrup AMR ultrasonic domestic water meters', says Mr Bhopale.

ECONOMIC WIN-WIN

The system was successfully installed with a hand-held drive-by solution. The water meters log daily consumption data and events such as leaks, tampering and reverse flow for 460 days, and monthly data for 36 months.

Due to it being a small council, Maharashtra state water supply does not have trained staff for meter reading activity. Therefore, Kamstrup provide practical training for local council staff. The training teaches staff how to read meters using a USB meter reader, transfer data to a local computer by using reading software, export reading data to their billing software, and prepare bills on a monthly basis for individual consumer.

TESTIMONIALS

Commenting at the time of implementation, Mr. B. K. Wankhede noted that: 'The system definitely delivers as expected. By using ultrasonic meters we will get accuracy and be able to save water. That awareness goes back to the consumer. What we deliver is what will be counted. The economy issue is therefore important for us and the consumer. Kamstrup gives us awareness of new technology and helps us achieve efficiency at the water plant and reduce non-revenue water.'

Mr. K. R. Otari, Sub-Divisional Engineer at Maharashtra Jeevan Pradhikaran, said: 'By getting a Kamstrup water meter, consumers experience tangible improvement of their daily life. During bulk supply, consumers had to wait from 6 AM till night, not knowing when the water will come on or how much there will be. This was always an unknown. Now consumers don't need to store their water and don't need to waste water.'

Case study 8

Real-life innovation

Kamstrup / Skanderborg Utility (Denmark)

A development project in close collaboration with Skanderborg Utility has provided new knowledge on the sound and noise generated by leaks and breaches in a distribution network. This has formed the basis for new and innovative leak detection in Kamstrup's newest water meter, which Skanderborg Utility expects will be of considerable value.

It started with an idea. Could the ultrasonic signal in Kamstrup's water meter also be used to measure acoustic changes, and in that way identify leaks in pipelines? The idea was good in theory – but it had to be tested in real life in order to ensure that it would generate value for the utilities.

'Our approach to product development has always been that a product remains unfinished until we know that it works for our customers. Innovation only becomes valuable once it has been tested in the real world,' explains Stig Knudsen, Product Manager at Kamstrup.

THE VALUE OF COLLABORATION

Skanderborg Utility was happy to be a part of the project, and Kamstrup applied to the Foundation for Development of Technology in the Danish Water Sector (Vandsektorens Teknologiudviklingsfond; VTU) fund to run it as a development project, which also included Børkop Water Utility. The agreement was settled and the project kicked-off in July 2015.

'A requirement for VTU projects is that water utilities and other industry players work together, so this was a great way to organise the work,' says Stig Knudsen. Skanderborg Utility has a strong focus on collaboration and innovation, which has also led to the establishment of the AquaGlobe innovation centre, housed in Skanderborg Utility, with the mission of supporting the development of Danish water technology.

There was also an substantial short-term reward in participating in the project: 'In the short term, it is all about spending fewer hours tracking down leaks. We have had cases where we had to listen through a large area over the course of several weeks. We would of course like to minimise that time,' says Jens Ravn Knudsen, Head of Asset Management at Skanderborg Utility.

With the new meter (see Figure CS8.1) it is easier for Skanderborg Utility to detect leaks in the distribution network. The meter picks up any sound changes enabling the utility to narrow down the area they need to go out and investigate, based on where the sound is loudest. 'The faster we can narrow

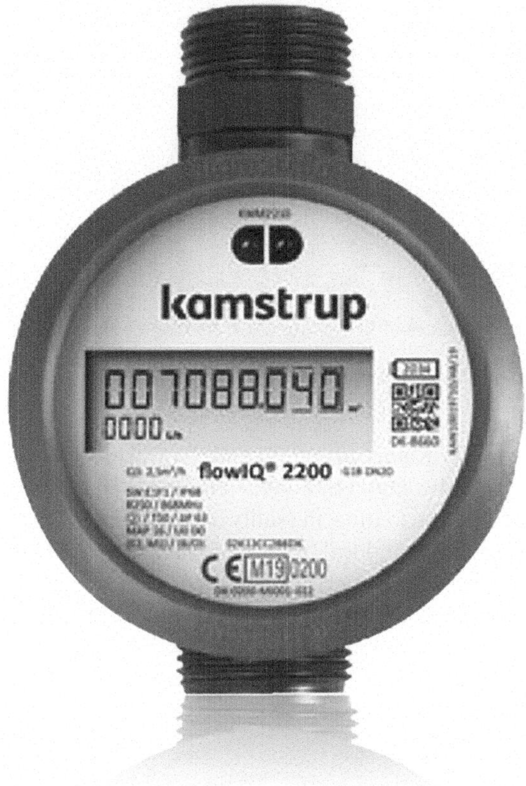

Figure CS8.1 A flowIQ 2200 flow meter.

down the area in which the leak might be, the quicker and cheaper we can rectify the damage instead of spending time and money on listening to an entire zone.'

WHAT DOES A HOLE SOUND LIKE?

Kamstrup started with a sandbox setup in the company's own back garden where they could test a string of prototypes. Here, the development engineers from Kamstrup and Frank Nielsen, Operational Assistant from Skanderborg Utility, drilled holes in test pipes – and listened. 'We spent a lot of time filtering the sound – recognising and eliminating noise sources such as traffic, a washing machine, etc.,' explains Stig Knudsen, continuing: 'When you have someone with years of operational experience and that mindset working side by side with an engineer with a PhD it gets really nerdy – and very productive.'

Prototypes were subsequently set up in different places around Skanderborg Utility's distribution network – either as part of planned renovations or in areas with suspected leakages that the prototypes then helped find.

First and foremost, Skanderborg Utility ensured that the meter was tested with the correct noise sources but another, equally valuable, contribution was their knowledge of sound and noise in the network. This was a completely new area for Kamstrup: what sound does a hole make? 'We had to learn what a leak sounds like', explains Stig Knudsen.

Figure CS8.2 Analysing leak holes in pipe samples.

Listening meetings were set up for Kamstrup and Skanderborg Utility to discuss just that, as the sound of a drilled hole is very different from the sound of a breach or leak. 'Based on our experience, we knew that drilling a clean hole in a pipe wasn't true to life – in reality a breach is usually a tear in the pipe, which creates a more whistling sound than a round hole does', explains Frank Nielsen (Figure CS8.2).

Skanderborg Utility therefore provided Kamstrup with real pieces of piping with leak holes to investigate and test in order to learn how the type of pipe, material, composition, pressure, etc. might affect the sound of a leak. This work was crucial to the development of the new meter, explains Stig Knudsen: 'The meter listens all day but needs to have a baseline. But when we started out, we didn't know what to look for. Today we know what a leak sounds like.'

LONG-TERM FOCUS ON ASSET MANAGEMENT

Skanderborg Utility's long-term focus is not the potential for reducing the utility's water loss, which is already less than 6%, following a massive effort on water loss reduction in 2010. Instead, asset management is their primary focus area (Figure CS8.3).

Figure CS8.3 Leak repair activity.

Jens F. Bastrup, CEO, explains: 'For us, the greatest value will come from being at the leading edge of asset management. This will allow us to safeguard our assets and spend our money in the right places and at the right times. That is the long-term goal. If we can use data to identify the signs of an impending breach, we will essentially be able to replace the exact pipe that needs replacing – the day before it breaks.'

Case study 9

Enabling the digital journey for a water utility

Elio Arniella, Smart Water Analytics (USA)

INTRODUCTION

Water utilities have traditionally used different methods to measure the volume and value of water at various phases of the water cycle. Many water utilities perform an annual audit using guidance and tools from sources, such as the International Water Association (IWA) and the American Water Works Association (AWWA). These tools and methodologies have been very useful in assisting water service providers in improving efficiency and sustainability. They have also increased awareness of the importance of conservation of resources and of reducing non-revenue water (NRW).

With new innovations in smart water infrastructure technologies (SWIT), water utilities can advance their water auditing and balancing methodologies by incorporating district metered areas (DMAs) with remote monitoring and automating of data analytics and reporting. These innovative tools allow utility operators to have the most recent and up-to-date system information and to identify and anticipate deficiencies in a pro-active manner. Coupled with the power of the cloud and advances in machine learning, SWIT can be extended to support an 'optimised control' to reduce NRW in real time.

The Case Study presented here concerns a water utility that changed its business model from one that made a once-a-year water audit to one that produces a daily and monthly Smart Water Balance (SWB). The SWB is a process driven by SWIT and information generated by the process itself, improved incrementally with knowledge acquired through its phased progression, change management, capacity building, technical support, proper funding and fact-based operational policies. The SWB process generates a common database of uncontested facts that can be used for planning and decision making. It is the reality check of where the utility is now in its digital journey and in accomplishing the mission and vision established in the business model.

THE GRAND BAHAMA UTILITY COMPANY

The Grand Bahama Utility Company (GBUC) is the second largest water utility in the Bahamas with 12 000 customers. The island's water supply is 100% from groundwater. The GBUC 2018 Water Master Plan laid out the road map for establishing a new business model that included modernising the water utility and improving its efficiency. The Master Plan recommended a short-term Phase 1 capital programme that included: building smart tools like GIS and a hydraulic model, universal smart metering with AMR/AMI, SCADA, district metering, pressure management, and integration/automation of information management systems.

Early in 2019, the GIS and hydraulic models were completed and used in the design of the Phase 1 implementation programme. After design and procurement of equipment and materials needed for the Phase 1 initiatives, the GBUC staff started the implementation of the programme components. Unfortunately, the programme was abruptly interrupted when Hurricane Dorian struck Grand Bahama Island from August 31 to September 2, 2019. The Category 5 hurricane with winds exceeding 185 mph and wind gusts reaching 220 mph caused an unprecedented storm surge of 20 feet (c.6.1 m). It also stalled over Grand Bahama Island for over 36 hrs. The storm was devastating and, by far, exceeded all benchmarks for damage and fatalities from previous storms in the region. The hurricane damaged most of the above ground water infrastructure and contaminated many its 250 production wells with salt water and high levels of dissolved solids (TDS).

After a speedy and diligent recovery, in January 2020 the GBUC was able to continue with the implementation phase of the programme by installing 18 flow and pressure monitoring stations in key areas of the systems, and establishing six DMAs. The data from the DMAs is transmitted every hour to the cloud using cellular technology. Initially, GBUC staff were able to log in to the vendor's website and download the data to an Excel spreadsheet. In April 2020, the system was automated using the cloud-based SWITLink™ software by Smart Water Analytics, where daily and monthly reports are automatically generated with key performance indicators (KPIs) to allow water operators to properly track water distribution and obtain daily and monthly electronic reports from each DMA. SWITLink™ generates five automated dashboards that provide data analytics regarding: (1) water treatment plant production; (2) daily DMA supply; (3) monthly DMA supply, consumption and NRW analytics; (4)

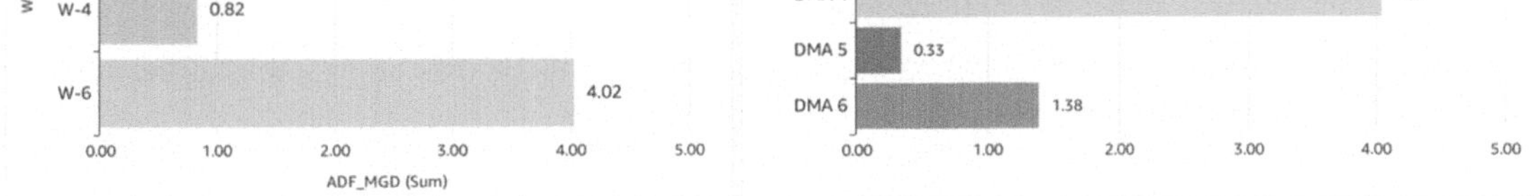

WTP Production and Pressure

WTP	Jul 27, 2020 ADF_GPM	ADF_MGD	AVG_PSI	MAX_PSI	MIN_PSI
W-1	2,056	2.96	41.39	46.45	32.17
W-3	889	1.28	34.99	37.05	20.01
W-4	572	0.82	23.93	28.18	15.10
W-6	2,794	4.02	43.50	45.45	12.12

DMA Daily Flow

DMA	DMA_ID	Jul 27, 2020 Customers	ADF_GPD	ADF_%	MAX_GPD	MIN_GPD
DMA 1	East	420	109,322	1.20	597,133	-199,806
DMA 2	Churchill	1,250	971,986	10.70	1,320,549	296,458
DMA 3	Lucaya	2,100	2,255,124	24.82	3,103,152	846,197
DMA 4	Freeport	5,691	4,035,715	44.41	4,699,199	2,229,782
DMA 5	Shipyard	1	331,686	3.65	942,088	-1,438
DMA 6	West End	2,050	1,383,640	15.23	1,966,171	585,989

Figure CS9.1 Example of the summary page of the GBUC SWITLink™ output.

BOX CS9.1. SWITLINK™

SWITLink™ is a scalable platform that connects to smart meters through open APIs (application programming interfaces) to retrieve, collect, store, calculate KPIs, and forecast future flows to help operators plan and optimise for minimum NRW. SWITLink™ is template driven to allow implementation in days.

The core components of SWITLink™ – integration with smart readers/SCADA, secure cloud storage, cloud analytics and machine learning – can also be leveraged to solve problems for wastewater and asset management.

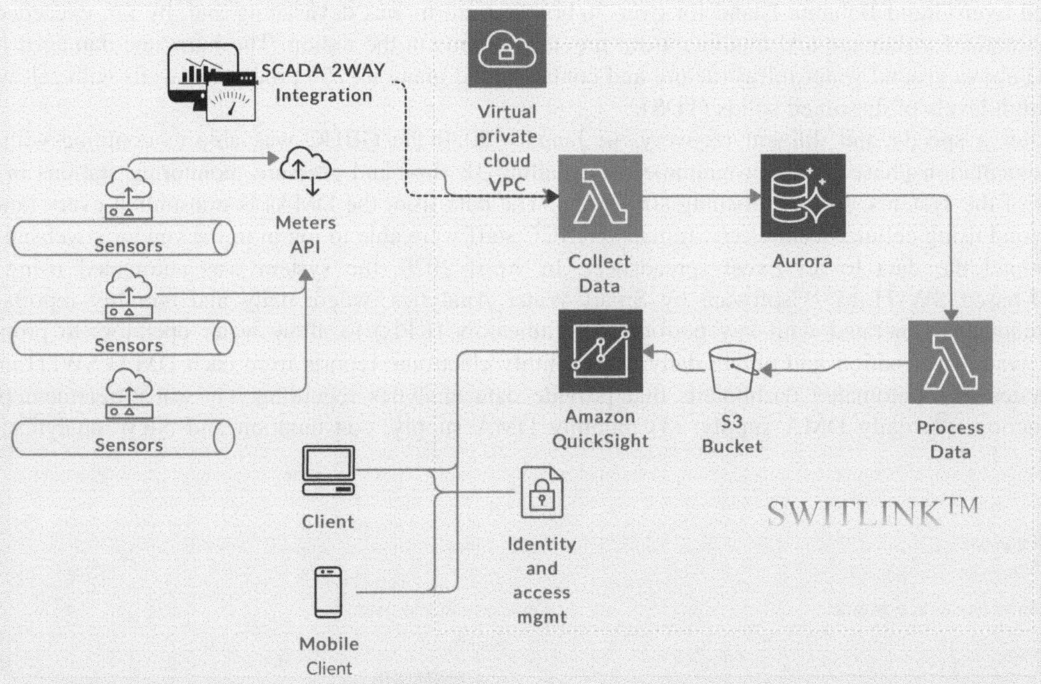

energy and chemical usage, and (5) a billing summary. Figure CS9.1 shows an example of the summary included in the daily report produced by the SWITLink™ software.

NEXT STEP

Figure CS9.2 summarises the GBUC digital journey as it transforms from manual reporting to cloud-based automation and beyond. As shown in Figure CS9.2, the next phase of the project is scheduled to be completed in the second quarter of 2021. In Phase 2, the GBUC SWITLink™ application will be enhanced to integrate new AMR/AMI customer water meter dashboards, and SCADA data to provide a 360° view of numerous KPIs regarding production, losses, NRW analytics

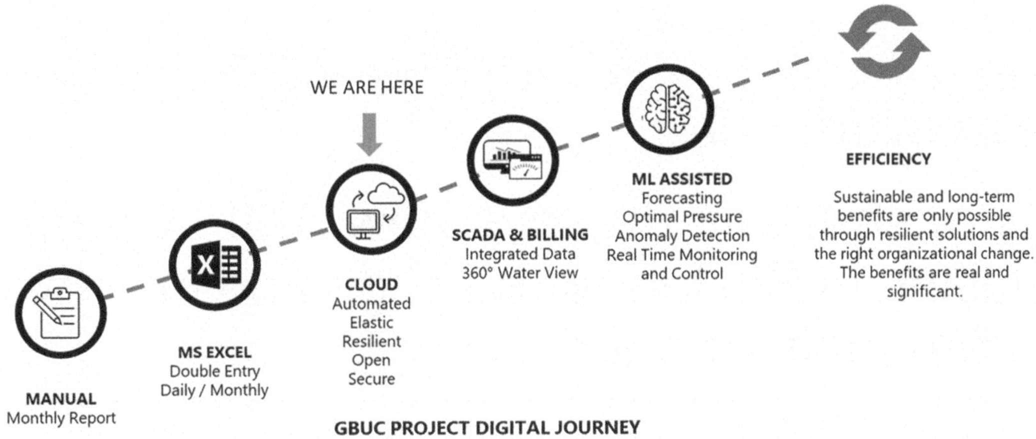

Figure CS9.2 The GBUC project digital journey.

and components, energy and chemicals, and billing data. The final phase is to incorporate predictive machine learning algorithms to control pressure and react timely to leaks and other system imbalances and deficiencies.

CONCLUSIONS

Water service providers worldwide are transitioning from conventional operations to automated mechanisation using random types of SWIT and innovative tools. One of the possible challenges faced by water utilities is not having a proper business model that lays out the requirements and commitments for the integration of SWIT and other innovative tools. There are also institutional challenges that hinder their ability to move forward in leveraging the full benefits of SWIT in a cost-effective manner. The GBUC Case Study presents an example of a utility that, despite having faced significant obstacles such as Hurricane Dorian and the Covid-19 pandemic, are moving forward in the implementation of a new business model. The new business model is facilitating the utility's digital conversion and is helping GBUC to improve its resiliency to face future storm events, obtaining the data needed to implement NRW reduction initiatives, and increasing long-term efficiency and profitability.

Case study 10

City of sheridan deploys new AMI system, resulting in Big labour savings and improved system maintenance

Mueller Systems (USA)

SUMMARY

The City of Sheridan, Wyoming, sought to implement a turn-key advanced metering infrastructure (AMI) solution that would not only help achieve accurate meter reading and billing, but also enhance customer relations. The rollout of the Mueller Mi.Net® AMI system on more than 10 000 service connections resulted in an estimated 65% drop in labour hours previously spent on servicing water meters and an improved daily read of 99.6%.

INTRODUCTION

Located in the state of Wyoming, USA, the City of Sheridan has a population of approximately 17 500 residents. Thwarted by aging meters installed almost 20 years ago, the City researched a smart meter system that would provide improved system reliability and future operational efficiencies. To achieve this goal, the City migrated from its existing automatic meter reading (AMR) system to a more powerful and flexible AMI system.

'One of the many things we want to achieve is to serve as an efficient unit that is equipped with the ability to carry out meter reading instantaneously at any time of the day, and be able to identify leaks in the system', said Ken Hirschman, Utility Maintenance Superintendent for the City of Sheridan. 'We also wanted to upgrade to smart water meters that are less mechanical and with less moving parts – effectively enabling us to get rid of mechanical meters and not be stuck with a load of inventory and a meter rebuilding process,' said Ken.

The City found the perfect solution to meet their goals in Mueller Systems' two-way Mi.Net® AMI network system, with its functionality and cost-effectiveness, amongst other attributes.

The Mi.Net system is a communication network that fully automates the meter-reading-to billing process by linking meters in a single, highly efficient data network. This provides the backbone for advanced metering technology as well as sensors for continuous monitoring of the entire water distribution system.

Figure CS10.1 Downtown Sheridan.

OPERATIONAL EFFICIENCY

The Mi.Net system has saved the City valuable labour hours, which are now deployed to other vital areas of their water and sewer systems. The City reports an average of 2300 staff hours saved annually on meter reading, work orders and turn offs. The utility maintenance team also calculated approximately 65% of their time saved on meter-related work, as they can now get instantaneous reads without having to get in a vehicle.

Over time, the City's utility maintenance team transitioned its dedicated person for meter reading to an operational role where he can contribute to other parts of the water and sewer systems, not just meters. These additional staff hours allow the City to accomplish more important activities such as cleaning and flushing sewer mains, saving the City an estimated $50 000 per year on sewer backup claims. Additionally, the City replaced six fire hydrants that resulted in increased system reliability on hydrants that were previously

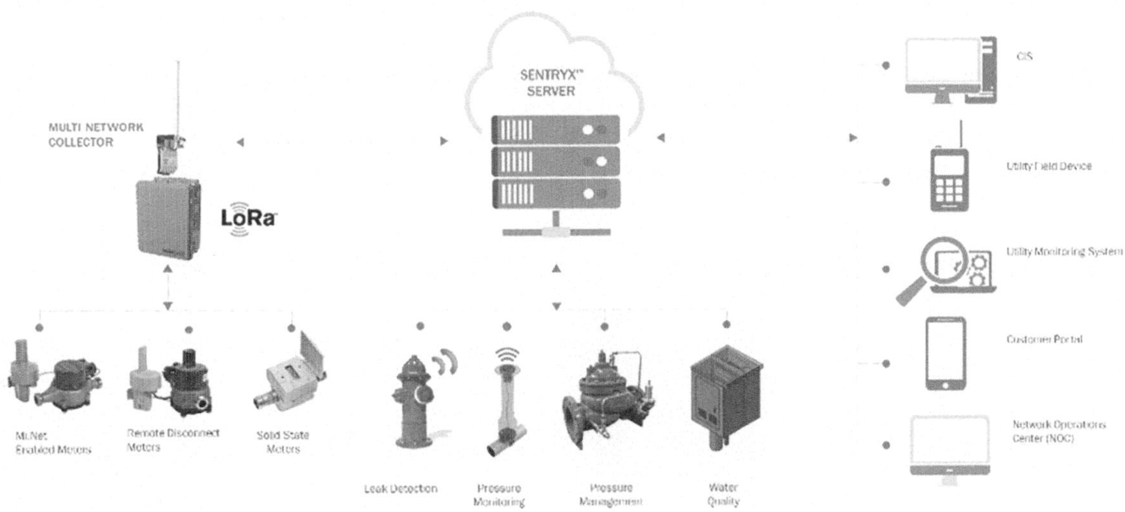

Figure CS10.2 Mueller Mi.Net® system architecture.

non-functioning or had low flow. 'We are able to serve our city and community much better than before. In a year, we saw our fire hydrant maintenance jump over 800 hours – something that we were never able to do; we had an extra 500 hours in sanitary sewer maintenance that was also unprecedented. Overall, our system maintenance has really improved. There have been less sewer plugs and claims against the city, and more time for replacing fire hydrants to safeguard our residents,' Ken shared.

The Mi.Net system integrates with Mueller Systems' remote disconnect meter (420 RDM), which, by accessing account information through the Mi.Net AMI user interface, allows meters to be turned on or off. The Mueller 420 RDM can be turned on or off from the office in under 18 seconds, saving valuable labour hours.

ON-DEMAND READ

The Mi.Net radio transmitter captures a read from each meter every hour. Each night, the node uploads all 24-hour reads to the Mi.Net servers and into the City's user interface. Should utility staff need a read urgently, a reading may be requested on-demand; the reading request is sent to the meter and read in real-time, feeding back to the operator within seconds. 'With the Mi.Net AMI system, I can do a quick water audit at any time of the day. The meter reading tells me how much water we have metered that day. Subsequently, I can look at our SCADA system which has our water treatment plant discharge flow, to make sure that they are fairly accurate,' said Ken.

The Mi.Net system has allowed the customer service team to remedy an issue that was formerly in the hands of another City division. 'If there are readings that did not pull through during the billing process, or if a customer suspects a leak in their service line, we do not automatically need to roll out a maintenance truck to investigate. We are able to immediately access the readings and usages before and after the date in question and help identify what is needed. With information available at daily and even hourly intervals, we can offer customers fully substantiated information to help them know with confidence how their home systems are functioning,' said Cathy Wright Bare, Customer Service Specialist for the City of Sheridan. It is estimated that the customer service team has been saving approximately 800 hours per year making use of the Mi.Net system's capabilities and efficiencies. This allows the members of the team to reallocate resources to other customer service endeavours.

From a water conservation perspective, the ability to capture hourly and on-demand data allows issues such as unauthorised domestic water use and network leaks to be addressed in a timely manner. For the City, a reduction in wholesale water costs and energy required for pumping during production and distribution can be realised; for customers, the self-service functionality empowers them to track their usage pattern, identify leaks and maximise water consumption efficiency.

Figure CS10.3 Mueller Systems' 420 Series Remote Disconnect Meter (RDM) with ME-8 Register.

Figure CS10.4 Mueller Network Operation Center (NOC) in Atlanta, Georgia.

AROUND THE CLOCK SYSTEM MANAGEMENT

All Mi.Net systems are continuously monitored by network engineers in a state-of-the-art Network Operations Center (NOC). The NOC employs software to analyse all communications and identify patterns, trends and conditions, allowing engineers to respond in real time and adjust remotely when required. 'The other great thing that we really liked about the Mi.Net system is the hosted option – they have the NOC which monitors our system as well as we do. If there is any issue, we do not have to have any extra personnel to manage that: NOC does it all. The team at Mueller Systems really demonstrate their willingness to be successful for this project and the community,' Ken added.

SAFE AND RELIABLE

Whilst designing a solution for this deployment, Mueller Systems wanted to not only focus on helping the City maximise their investment in an AMI system by providing a true two-way network that allows upgrades, but also on one of the most important factors – network security. From end to end, the Mi.Net system provides multiple layers of security to prevent hacking, impersonation and other network security threats. Security mechanisms and protocols are designed into the system to enable hosting of all software in a secure environment and providing round the clock monitoring.

Additionally, the Mi.Net system architecture employs LoRa™ based applications to establish a reliable and secure layer that can prevent accidental loss and/or interception of customer data. This provides unmatched accuracy, distance, bandwidth, throughput and resistance to interference in the unlicensed industrial, scientific and medical (ISM) band between 902 and 928 MHz. The use of LoRa radio frequency modulation offers high-power transmissions and an increased range over traditional systems while having lower battery usage. The proof is in the numbers. Since deployment, Sheridan's Mi.Net system is averaging a 99.6% read rate.

Mueller Systems continues to work alongside the City of Sheridan to empower them to do more with less and accelerate its journey to becoming a smart city of the future with the power of Mi.Net AMI system.

IWA Publishing's authorised EU representative for General Product Safety Regulations is Diane D'Arras, 15 rue Duret, 75116 Paris, France, e-mail: safety@iwap.co.uk.

Printed and bound by CPI Group (UK) Ltd, Croydon, CR0 4YY
12/05/2026
02108869-0019